Soil Contamination by Heavy Metals and Metalloids

Soil Contamination by Heavy Metals and Metalloids

Editor

Dionisios Gasparatos

MDPI • Basel • Beijing • Wuhan • Barcelona • Belgrade • Manchester • Tokyo • Cluj • Tianjin

Editor
Dionisios Gasparatos
Agricultural University
of Athens
Greece

Editorial Office
MDPI
St. Alban-Anlage 66
4052 Basel, Switzerland

This is a reprint of articles from the Special Issue published online in the open access journal *Environments* (ISSN 2076-3298) (available at: https://www.mdpi.com/journal/environments/special_issues/Soil_Contamination).

For citation purposes, cite each article independently as indicated on the article page online and as indicated below:

LastName, A.A.; LastName, B.B.; LastName, C.C. Article Title. *Journal Name* **Year**, *Volume Number*, Page Range.

ISBN 978-3-0365-3545-6 (Hbk)
ISBN 978-3-0365-3546-3 (PDF)

Cover image courtesy of Dionisios Gasparatos and Kristy Kokkinou

© 2022 by the authors. Articles in this book are Open Access and distributed under the Creative Commons Attribution (CC BY) license, which allows users to download, copy and build upon published articles, as long as the author and publisher are properly credited, which ensures maximum dissemination and a wider impact of our publications.

The book as a whole is distributed by MDPI under the terms and conditions of the Creative Commons license CC BY-NC-ND.

Contents

About the Editor vii

Dionisios Gasparatos
Soil Contamination by Heavy Metals and Metalloids
Reprinted from: *Environments* **2022**, *9*, 32, doi:10.3390/environments9030032 1

Juan Miguel Moreno-Alvarez, Rosa Orellana-Gallego and Maria Luisa Fernandez-Marcos
Potentially Toxic Elements in Urban Soils of Havana, Cuba
Reprinted from: *Environments* **2020**, *7*, 43, doi:10.3390/environments7060043 5

**Shamali De Silva, Trang Huynh, Andrew S. Ball, Demidu V. Indrapala and
Suzie M. Reichman**
Measuring Soil Metal Bioavailability in Roadside Soils of Different Ages
Reprinted from: *Environments* **2020**, *7*, 91, doi:10.3390/environments7100091 23

Ioannis Zafeiriou, Dionisios Gasparatos and Ioannis Massas
Adsorption/Desorption Patterns of Selenium for Acid and Alkaline Soils of
Xerothermic Environments
Reprinted from: *Environments* **2020**, *7*, 72, doi:10.3390/environments7100072 37

**Liqiang Cui, Lianqing Li, Rongjun Bian, Jinlong Yan, Guixiang Quan, Yuming Liu,
James A. Ippolito and Hui Wang**
Short- and Long-Term Biochar Cadmium and Lead Immobilization Mechanisms
Reprinted from: *Environments* **2020**, *7*, 53, doi:10.3390/environments7070053 49

João Antonangelo and Hailin Zhang
Influence of Biochar Derived Nitrogen on Cadmium Removal by Ryegrass in a
Contaminated Soil
Reprinted from: *Environments* **2021**, *8*, 11, doi:10.3390/environments8020011 65

**Muhammad Zafar-ul-Hye, Muhammad Naeem, Subhan Danish, Shah Fahad, Rahul Datta,
Mazhar Abbas, Ashfaq Ahmad Rahi, Martin Brtnicky, Jiří Holátko, Zahid Hassan Tarar and
Muhammad Nasir**
Alleviation of Cadmium Adverse Effects by Improving Nutrients Uptake in Bitter Gourd
through Cadmium Tolerant Rhizobacteria
Reprinted from: *Environments* **2020**, *7*, 54, doi:10.3390/environments7080054 77

About the Editor

Dionisios Gasparatos is Associate Professor of Soil Science in the Laboratory of Soil Science and Agricultural Chemistry at the Agricultural University of Athens, Greece. He received his MSc in Applied Environmental Geology from The National and Kapodistrian University of Athens, and his PhD in Soil Science from Agricultural University of Athens. His teaching and research focuses on pedology, soil fertility, plant nutrition, soil mineralogy, and chemistry with environmental applications. He is author of more than 60 papers and chapters published in high impact international journals and books, including Environmental Chemistry Letters, Catena, Chemosphere, Journal of Hazardous Materials, Journal of Plant Nutrition, and Soil Science with over 750 citations, and is naturally a reviewer and editorial board member of several international journals as well as an editor of textbooks for graduate students. His scientific interests are related to a wide variety of soil problems including genesis of redoximorphic features in saturated soils, amendment effects on potential toxic elements, integrated soil nutrient management practices, soil–plant relations, phosphorus dynamics, and key threats to soil. Website: https://www.researchgate.net/profile/Dionisios-Gasparatos.

Editorial

Soil Contamination by Heavy Metals and Metalloids

Dionisios Gasparatos

Laboratory of Soil Science and Agricultural Chemistry, Agricultural University of Athens, 75 Iera Odos Street, 118 55 Athens, Greece; gasparatos@aua.gr

Citation: Gasparatos, D. Soil Contamination by Heavy Metals and Metalloids. *Environments* **2022**, *9*, 32. https://doi.org/10.3390/environments9030032

Received: 21 February 2022
Accepted: 1 March 2022
Published: 5 March 2022

Publisher's Note: MDPI stays neutral with regard to jurisdictional claims in published maps and institutional affiliations.

Copyright: © 2022 by the author. Licensee MDPI, Basel, Switzerland. This article is an open access article distributed under the terms and conditions of the Creative Commons Attribution (CC BY) license (https://creativecommons.org/licenses/by/4.0/).

Soils are central to life on Earth because they provide food, clean water, and air due to their filtering capacity; raw materials; habitats for living organisms; and climate resilience via carbon sequestration, therefore supporting a variety of ecosystem services [1]. Despite this, soil's life-sustaining functions have been underestimated until recently, and few people seem to be aware of them. It is now clear that soil plays an active role in maintaining life and that life is unable to exist without it. The insights about soils that have been gleaned over the last few decades have revealed that these complex systems are fragile, scarce, non-renewable, threatened resources as well as a crucial link between local and global environmental issues. Healthy soils are critical to achieving the Sustainable Development Goals (SDGs) outlined in the United Nations and the Europe Green Deal strategy [2]. In the Europe context, it is estimated that between 60 and 70% of EU soils are unhealthy due to erosion, contamination, compaction, carbon depletion, biodiversity loss, and sealing.

Soil contamination has been identified and as one of the main threats to soil, inducing the degradation of global soils and driving long-term losses of the ecosystem services that they provide. As a result of human activities, the amount of soil contamination caused by heavy metal(loid)s has severely increased over the last few decades and has become a worldwide environmental issue that has attracted considerable public attention [3,4]. Soil contamination changes biota composition and affects water, air, and food quality, degrading human health. Although many research efforts have highlighted how soil contamination is a global threat, providing an overview of the importance of healthy soil, there is still a great need for additional information from different regions around the world, and concrete strategies, which can be implemented to address the causes and impacts of this major threat, urgently need to be developed.

In this context, this Special Issue was launched with the scope of bringing together articles presenting the development of novel science-based methods and applications that enhance the remediation of contaminated soil by focusing on (1) the identification of the main sources of soil contamination caused by heavy metal(loid)s (HM)/potentially toxic elements (PTEs) in different soil types, (2) the chemistry, speciation, potential mobility, and bioavailability of the contaminants that are commonly found in contaminated soils, (3) the assessment of the negative impacts and risks associated with HM/PTE-induced soil contamination on crop yields, soil biota, food security, and human health, (4) the available methods and strategies for monitoring, assessing, and remediating soils that have been contaminated by HM/PTEs, and (5) guidelines that include threshold values for HM/PTE levels in soils at national and regional levels.

HM/PTE contamination in anthropogenic urban soils constitute a major environmental problem, and therefore, urban soils are often subject to detailed risk-assessment and management studies [5,6]. Contamination is often assessed in terms of total concentrations, revealing possible soil enrichment due to heavy metals, and is used to establish regulatory guidelines for policy decisions. However, various authors have recognized that it is the bioavailability of metals that determines their fate and behavior in the environment [7,8]. In this respect, the article of Moreno-Alvarez et al. [9] focuses on the pseudo-total, -available, and -acid-oxalate concentrations that are extractable from urban soils of Havana, Cuba.

They observe that the pseudo-total concentrations were generally higher than the average values for the world's soils that they were similar to those published for urban soils. Moreover, the study shows that Fe, Ti, V, Ni, Cr, and Co were mainly of lithological origin, whereas the contribution of anthropogenic sources related to industrial activities, fuel combustion, and the application of organic amendments to soil were connected to Cr, As, Hg, Pb, Zn, Cd, and Cu levels. The toxicity limits for the bioavailability of Cd, Ni, Mn, and Pb were exceeded by 14%, 10%, 39%, and 56% of samples, respectively; therefore, the authors argue that guidelines underlining the safety limits for the better management of the urban agriculture activities should be introduced.

As they comprise a significant part of urban ecosystems, roadside soils have attracted the interest of researchers for at least four decades [10]. Roadside soils are greatly affected by man-made changes that affect their physicochemical and biological properties, and many studies have pointed out that the metal concentrations in these soils depend on traffic intensity. In that context, De Silva et al. [11] deal with the bioavailability of metals aged in roadside soils in situ. Using soil samples collected from sites representing three road of different ages (new, medium, and old roads) and different techniques (diffusive gradients in thin films technique, soil water extraction, $CaCl_2$ extraction, total metal concentrations, and optimized linear models), the study provides insights into the kinetics of the long-term metal aging that causes metal bioavailability to decrease over time.

Although some PTEs such as Selenium (Se) are considered essential micronutrients at low concentrations, they can be hazardous to human health at concentrations exceeding tolerable doses [12]. Given this constrain, Zafeiriou et al. [13] examined the adsorption and desorption processes of Se(IV) that had been freshly added to acid and alkaline soils, aiming to describe the Se's geochemistry in terms of its bioavailability and contamination risk via leaching. The acidic soils adsorbed significantly higher amounts of the added Se(IV) than the alkaline soils did, and the alkaline soils desorbed more Se. Taking into consideration that biofortification through plant uptake is also crop/plant-dependent, the application of Se(IV) to agricultural soils should be site-specific, as Se poses a high leaching hazard risk in alkaline soils with low concentrations of metal oxides, while low Se availability may result in acidic soils with a high metal oxide content.

Several soil remediation strategies have been used among the scientific community in the last twenty years [14]. The application of low-cost and eco-friendly materials as immobilizing agents in HM/PTEs-contaminated soils has gained significant interest, and there is now an urgent need to understand their functionality in contaminated soils [15]. The extensive application of biochar, which acts as a contaminant scavenger, has received significant consideration for HM/PTE-contaminated soil remediation [16,17].

In this respect, the article by Cui et al. [18] focuses on the effect of wheat straw biochar on soil Cd and Pb bioavailability, uptake, and translocation by rice in a contaminated paddy soil. They observed that biochar application reduced Cd (16.1–84.1%) and Pb (4.1–40.0%) transfer from root to rice grain, simultaneously improving physicochemical (moisture content, pH, organic matter) and biological (enzymatic activity and microbial community structure) properties of the soil. According to the authors, understanding the role of biochar at the atomic level is critical to unraveling the mechanisms by which biochar stabilizes in situ heavy metals over longer time periods and in different soil types.

Phytoextraction, a strategy that uses plants to accumulate HM/PTEs in the biomass, is an alternative approach to restore contaminated soils and that has several limitations impeding its use in commercial applications [19,20]. As a practical example of this phytoremediation approach, Antonangelo and Zhang [21] dealt with the influence of nitrogen (N) added via biochars from different feedstocks on the cadmium (Cd) removal ability by ryegrass. Interestingly the authors applied different biochar doses to shows how N accumulation increases as a function of the biochar application rate, and this increase contributed to higher ryegrass yield and Cd accumulation.

Microbially-assisted phytoremediation is an innovative strategy that is based on the use of plant growth-promoting rhizobacteria (PGPRs) to enhance the phytoremediation efficiency [22]. In this respect, the article by Zafar-ul-Hye [23] focuses on the effect of *Stenotrophomonas maltophilia* and *Agrobacterium fabrum* to improve nutrient uptake and to alleviate the adverse effects of Cd in bitter gourd. The results showed that the treatment of *A. fabrum* combined with NPK fertilizers showed an increase in the number of bitter gourds per plant (34% and 68%); fruit length (19% and 29%); bitter gourd yield (26.5% and 21.1%); and N (48% and 56%) and K (72% and 55%) concentrations of the control grown under different soil cadmium concentration (2 and 5 mg kg^{-1} soil). The study concluded that *A. fabrum* is more effective that *S. maltophilia* in alleviating Cd-induced stress in bitter gourd.

Despite the increasing awareness of soil degradation due to contamination, as pointed out in this Special Issue, much more research is needed to better understand the behavior of HM/PTEs in soils with different pedoclimatic conditions, their interactions with soil components, and the development innovative and efficient remediation procedures for the recovery of contaminated soils.

Acknowledgments: I would like to express my sincere gratitude to the authors who contributed to this Special Issue, to the reviewers for their valuable assistance as well as to the staff of MDPI for their efforts to complete and publish this issue.

Conflicts of Interest: The author declares no conflict of interest.

References

1. Baveye, P.C.; Baveye, J.; Gowdy, J. Soil "ecosystem" services and natural capital: Critical appraisal of research on uncertain ground. *Front. Environ. Sci.* **2016**, *4*, 41. [CrossRef]
2. Lal, R.; Bouma, J.; Brevik, E.; Dawson, L.; Field, D.J.; Glaser, B.; Hatano, R.; Hartemink, A.E.; Kosaki, T.; Lascelles, B.; et al. Soils and sustainable development goals of the United Nations: An International Union of Soil Sciences perspective. *Geoderma Reg.* **2021**, *25*, e00398. [CrossRef]
3. Tóth, G.; Hermann, T.; Da Silva, M.R.; Montanarella, L. Heavy metals in agricultural soils of the European Union with implications for food safety. *Environ. Int.* **2016**, *88*, 299–309. [CrossRef] [PubMed]
4. Alengebawy, A.; Abdelkhalek, S.T.; Qureshi, S.R.; Wang, M.Q. Heavy metals and pesticides toxicity in agricultural soil and plants: Ecological risks and human health implications. *Toxics* **2021**, *9*, 42. [CrossRef]
5. Massas, I.; Gasparatos, D.; Ioannou, D.; Kalivas, D. Signs for secondary build up of heavy metals in soils at the periphery of Athens International Airport, Greece. *Environ. Sci. Poll. Res.* **2018**, *25*, 658–671. [CrossRef]
6. Herbón, C.; Barral, M.T.; Paradelo, R. Potentially toxic trace elements in the urban soils of Santiago de Compostela (Northwestern Spain). *Appl. Sci.* **2021**, *11*, 4211. [CrossRef]
7. Zafeiriou, I.; Gasparatos, D.; Kalyvas, G.; Ioannou, D.; Massas, I. Desorption of arsenic from calcareous mine affected soils by phosphate fertilizers application in relation to soil properties and As partitioning. *Soil Syst.* **2019**, *3*, 54. [CrossRef]
8. Antoniadis, V.; Shaheen, S.M.; Boersch, J.; Frohne, T.; Laing, G.D.; Rinklebe, J. Bioavailability and risk assessment of potentially toxic elements in garden edible vegetables and soils around a highly contaminated former mining area in Germany. *J. Environ. Manag.* **2017**, *186*, 192–200. [CrossRef]
9. Moreno-Alvarez, J.M.; Orellana-Gallego, R.; Fernandez-Marcos, M.L. Potentially Toxic Elements in Urban Soils of Havana, Cuba. *Environments* **2020**, *7*, 43. [CrossRef]
10. Werkenthin, M.; Kuge, B.; Wessolek, G. Metals in European roadside soil and soil solution—A review. *Environ. Pollut.* **2014**, *189*, 98–110. [CrossRef]
11. De Silva, S.; Huynh, T.; Ball, A.S.; Indrapala, D.V.; Reichman, S.M. Measuring Soil Metal Bioavailability in Roadside Soils of Different Ages. *Environments* **2020**, *7*, 91. [CrossRef]
12. Zafeiriou, I.; Gasparatos, D.; Ioannou, D.; Kalderis, D.; Massas, I. Selenium Biofortification of Lettuce Plants (*Lactuca sativa* L.) as Affected by Se Species, Se Rate, and a Biochar Co-Application in a Calcareous Soil. *Agronomy* **2022**, *12*, 131. [CrossRef]
13. Zafeiriou, I.; Gasparatos, D.; Massas, I. Adsorption/Desorption Patterns of Selenium for Acid and Alkaline Soils of Xerothermic Environments. *Environments* **2020**, *7*, 72. [CrossRef]
14. Kalyvas, G.; Gasparatos, D.; Liza, C.A.; Massas, I. Single and combined effect of chelating, reductive agents, and agroindustrial by-product treatments on As, Pb, and Zn mobility in a mine-affected soil over time. *Environ. Sci. Pollut. Res.* **2020**, *27*, 5536–5546. [CrossRef]
15. Palansooriya, K.N.; Shaheen, S.M.; Chen, S.S.; Tsang, D.C.W.; Hashimoto, Y.; Hou, D.; Bolan, N.S.; Rinklebe, J.; Ok, Y.S. Soil amendments for immobilization of potentially toxic elements in contaminated soils: A critical review. *Environ. Int.* **2020**, *134*, 105046. [CrossRef]
16. He, L.; Zhong, H.; Liu, G.; Dai, Z.; Brookes, P.C.; Xu, J. Remediation of heavy metal contaminated soils by biochar: Mechanisms, potential risks and applications in China. *Environ. Pollut.* **2019**, *252*, 846–855. [CrossRef]

17. Bilias, F.; Nikoli, T.; Kalderis, D.; Gasparatos, D. Towards a Soil Remediation Strategy Using Biochar: Effects on Soil Chemical Properties and Bioavailability of Potentially Toxic Elements. *Toxics* **2021**, *9*, 184. [CrossRef]
18. Cui, L.; Li, L.; Bian, R.; Yan, J.; Quan, G.; Liu, Y.; Ippolito, J.A.; Wang, H. Short- and Long-Term Biochar Cadmium and Lead Immobilization Mechanisms. *Environments* **2020**, *7*, 53. [CrossRef]
19. Kalyvas, G.; Tsitselis, G.; Gasparatos, D.; Massas, I. Efficacy of EDTA and Olive Mill Wastewater to Enhance As, Pb, and Zn Phytoextraction by *Pteris vittata* L. from a Soil Heavily Polluted by Mining Activities. *Sustainability* **2018**, *10*, 1962. [CrossRef]
20. Ashraf, S.; Ali, Q.; Zahir, Z.A.; Ashraf, S.; Asghar, H.N. Phytoremediation: Environmentally sustainable way for reclamation of heavy metal polluted soils. *Ecotoxicol. Environ. Saf.* **2019**, *174*, 714–727. [CrossRef]
21. Antonangelo, J.; Zhang, H. Influence of Biochar Derived Nitrogen on Cadmium Removal by Ryegrass in a Contaminated Soil. *Environments* **2021**, *8*, 11. [CrossRef]
22. Gul, I.; Manzoor, M.; Hashim, N.; Shah, G.M.; Waani, S.P.T.; Shahid, M.; Antoniadis, V.; Rinklebe, J.; Arshad, M. Challenges in Microbially and Chelate-Assisted Phytoextraction of Cadmium and Lead-A Review. *Environ. Pollut.* **2021**, *287*, 117667. [CrossRef] [PubMed]
23. Zafar-ul-Hye, M.; Naeem, M.; Danish, S.; Fahad, S.; Datta, R.; Abbas, M.; Rahi, A.A.; Brtnicky, M.; Holátko, J.; Tarar, Z.H.; et al. Alleviation of Cadmium Adverse Effects by Improving Nutrients Uptake in Bitter Gourd through Cadmium Tolerant Rhizobacteria. *Environments* **2020**, *7*, 54. [CrossRef]

Article

Potentially Toxic Elements in Urban Soils of Havana, Cuba

Juan Miguel Moreno-Alvarez [1], Rosa Orellana-Gallego [2,†] and Maria Luisa Fernandez-Marcos [1,3,*]

1. Department of Soil Science and Agricultural Chemistry, University of Santiago de Compostela, 27002 Lugo, Spain; juanm.morenoalvarez@gmail.com
2. Instituto de Investigaciones Fundamentales en Agricultura Tropical "Alejandro de Humboldt" (INIFAT), Havana 17200, Cuba; orellana@inifat.esihabana.cu
3. Institute of Agricultural Biodiversity and Rural Development, University of Santiago de Compostela, 27002 Lugo, Spain
* Correspondence: mluisa.fernandez@usc.es; Tel.: +34-982-823-119
† Deceased.

Received: 13 May 2020; Accepted: 5 June 2020; Published: 9 June 2020

Abstract: Urban soils are characterised by a strong anthropogenic influence. Potentially toxic elements were studied in various horizons of 35 urban soils in Havana, Cuba, classified as Urbic or Garbic Technosols. Pseudo-total, available, and acid-oxalate extractable concentrations were determined. The pseudo-total concentrations were generally higher than the average values for the world's soils but similar to those published for urban soils. In a few cases, very high values of copper or lead were found. Nickel and chromium concentrations exceeded the maximum allowable concentrations for agricultural soils in 22% and 12% of samples. Vanadium concentrations were always very high. There was minimum enrichment of most samples in Co, Mn, As, Cd, Cr, Cu, and Ni, but outliers reached moderate or significant enrichment. Enrichment was significant for V, while for Pb, Zn, and Hg the median values denoted moderate enrichment, but outliers reached significant enrichment in Zn and extremely high enrichment in Pb and Hg. The available elements amounted to between 0.07% of the pseudo-total vanadium and 30% lead and cadmium. The published toxicity limits for bioavailable Cd, Mn, Ni, and Pb were exceeded in 14%, 39%, 10%, and 56% of samples, respectively. The concentrations of pseudo-total total iron, cobalt, chromium, and nickel, and available cobalt, nickel and titanium were significantly lower in soils with gleyic properties (reducing conditions).

Keywords: PTE; anthropogenic soils; Technosols; trace elements; heavy metals; urban agriculture; heavy metal availability; enrichment factor; redox

---

1. Introduction

Soils in urban environments are strongly influenced by man, whose activity is based in these environments. The anthropogenic influence alters the processes of soil formation, often changing the direction of soil evolution. The human perturbation may vary in intensity, giving rise to a wide range of urban soils from quasi-natural to strongly disturbed soils. Human influence can interfere more or less intensely with the natural processes of soil formation, can provide exogenous materials, including pollutants, and can eventually build new soils by providing organic or mineral materials from which new processes of soil formation will start. Urban soils usually show a high vertical and horizontal variability, often contain artefacts, and are classified in the WRB [1] as Technosols.

On the other hand, urban soils provide considerable ecosystem services, such as supporting plant growth, including urban agriculture; taking part in biogeochemical cycles; contributing to the hydrological cycle; storing carbon and regulating greenhouse gases; modulating urban climate and air quality; contributing to urban biodiversity; supporting human activities; and intercepting and immobilizing or decomposing contaminants [2,3]. Therefore, having healthy urban soils is of paramount importance to enhance the quality of the urban environment and foster urban sustainable development.

The presence of high levels of trace elements and heavy metals is a major threat to the quality of urban soils, being of special concern in those involved in urban agriculture. A number of published articles studied potentially toxic elements (PTEs) in urban and peri-urban soils around the world [4–17]. Virtually all of them reported concentrations of heavy metals in urban soils higher than in natural soils.

The Cuban urbanization process is one of the oldest in the Americas. The urban population (8,637,568 inhabitants) constitutes 77% of the total residents of Cuba in 2018. Of this urban people, 2,131,480 inhabitants reside in the province of Havana [18]. One hundred percent of the province of Havana is urbanized. Given the size of the urban population, urban agriculture, which emerged in Cuba in 1987 as a programme of the Cuban Government, contributes significantly to local food self-sufficiency and to the country's food security [19]. Urban agriculture uses locally produced organic fertilizers with agro-ecological pest control and local seed production, avoiding the use of petrochemicals [19,20]. Along with urban agricultural soils, soils of parks and gardens, urban groves, and vacant areas constitute the urban soils of Havana.

This paper aims to study the concentrations and forms of PTEs in urban soils in the province of Havana, as a contribution to the knowledge of the quality of these soils, and to identify the possible origin of these elements. For such purposes, 35 urban soils were selected having a variety of uses (including urban agriculture) and intensity of human intervention.

2. Materials and Methods

2.1. Study Area

Havana is the province with the smallest area in Cuba, occupying 727.4 km^2. It is situated on the northern coast, surrounding the Havana Bay, where the main port of the country is located. The province has a population of 2.13 million inhabitants and a population density of 3000 inhabitants/km^2. One hundred percent of the province is urbanized. The climate is tropical with two distinct seasons: a wetter season (May–October) and a drier season (November–April). The monthly average temperature ranges between 22 and 27 °C, with an annual average of 25.7 °C. The annual average relative humidity is 80%, and the annual rainfall is 1240 mm. The geology of the area is mainly sedimentary limestone and ultramafic rocks, sometimes serpentinized.

Several industries are located in this province, which could well be a source of PTEs, including an oil refinery, metal processing industries, and power plants. The dense traffic of motor vehicles, some of them very old, could also contribute to the emissions.

The province consists of 15 districts, distributed in three areas: central, intermediate, and peripheral. Soils were selected for study in 6 of the 15 districts, 5 in the central area, the most densely populated, and 1 (Playa) in the intermediate area. The latter district has experienced a rapid socio-economic and touristic development in recent decades.

2.2. Soils

Thirty-five representative soils (Figure 1) were sampled in 6 of the 15 districts of the province of Havana, Cuba (Regla, San Miguel del Padrón, Plaza de la Revolución, 10 de Octubre, Playa, and Habana Vieja). The 35 soils have different uses: urban agriculture, parks and gardens, wooded areas, and vacant areas. The presence of PTEs in agricultural soils is of concern because of the risk of metal exportation to the food chain. In the case of parks and gardens, the main concern is related to the presence of children playing in these areas [15].

The agricultural soils were often built from soil materials transported from other locations. Some of them have a surface layer made of compost from various organic wastes. The urban agriculture system developed in these soils is known as organoponics [21,22]. Others only had manure and vermicompost applied.

The studied soils are moderately alkaline (pH = 8.27 ± 0.20) and usually rich in calcium carbonate ($CaCO_3$) in the surface horizon. The most frequent textures were clay, loam, or clay loam. According to the World Reference Base for Soil Resources [1], the selected soils were classified as Urbic or Garbic Technosols. Often the water table was near the soil surface, so that the soils had hydromorphic features. Twelve of the 35 soils studied showed gleyic properties. Gleyic properties mean current or past soil saturation with groundwater for a period that allows reducing conditions to occur [1].

Figure 1. Location of the studied soils in Havana, Cuba (map from Google Earth).

Several horizons were morphologically differentiated and sampled for each soil. Soil samples were air-dried and sieved (<2 mm) before analysis.

2.3. Analysis of PTEs

Thirteen heavy metals and trace elements were determined: arsenic (As), cadmium (Cd), cobalt (Co), chromium (Cr), copper (Cu), iron (Fe), manganese (Mn), nickel (Ni), lead (Pb), titanium (Ti), vanadium (V), zinc (Zn), and mercury (Hg). Pseudo-total and available elements were determined for each soil horizon (91 samples). PTEs associated with noncrystalline Al and Fe were determined only in surface horizons.

The determination of pseudo-total PTEs was performed by microwave-assisted digestion, with concentrated HNO_3 and HCl ($HCl:HNO_3$ ratio 1:3), according to the EPA 3051A method [23]. The metals were determined in the digestate by ICP-MS. The limits of detection, calculated as 3 times the standard deviation of blank values, in mg kg^{-1}, were: As, 0.09; Cd, 0.006; Co, 0.03; Cr, 4.6; Cu, 1.1; Fe, 8.3; Mn, 1.0; Ni, 0.32; Pb, 1.0; Ti, 2.0; V, 2.8; Zn, 6.1; and Hg, 0.01.

Available PTEs were extracted by Mehlich 3 reagent [24] and determined by ICP-MS.

PTEs associated with noncrystalline Al and Fe were extracted by ammonium oxalate/oxalic acid at pH 3 [25] and determined by ICP-MS.

2.4. Pollution Assessment

Two indices were used to assess pollution in the studied soils [13]. The enrichment factor (EF) compares the concentration of an element in the studied soil with the background concentration, both concentrations standardized against a reference element (Al, Fe, or Ti). We chose Fe as the reference element, so that EF was calculated as:

$$EF_X = (C_X/C_{Fe})_{soil}/(C_X/C_{Fe})_{background}$$

where X represents the element of interest; EF_X, the EF of element X; C_X, the total concentration of element X; C_{Fe}, the total concentration of iron (a lithogenic reference element); "soil" refers to the assessed soil, and "background" to the situation of natural unpolluted soils in the study area, taken as a reference. EF values below 2 indicate minimum enrichment; values between 2 and 5 denote moderate enrichment; significant enrichment is attributed to values between 5 and 20; values above 20 show very high enrichment; while values above 40 reveal extremely high enrichment [13].

The geoaccumulation index (I_{geo}) for a given element is calculated as:

$$I_{geo} = \log_2 (C_X/1.5\ B_X)$$

where C_X is the total concentration of element X; B_X is the background value for element X; and the factor 1.5 is used to minimize the effect of possible variations in the background due to lithological variations. According to I_{geo} values, the soils are classified into seven categories: <0 = practically unpolluted, 0–1 = unpolluted to moderately polluted, 1–2 = moderately polluted, 2–3 = moderately to strongly polluted, 3–4 = strongly polluted, 4–5 = strongly to extremely polluted, and >5 = extremely polluted [13].

As background concentrations to calculate EF and I_{geo}, we used the quality reference values (QRVs) established by Rodríguez Alfaro et al. [26] for Cuban soils:

$$EF_X = (C_X/C_{Fe})/(QRV_X/QRV_{Fe})$$

$$I_{geo} = \log_2 (C_X/1.5\ QRV_X)$$

QRV_{Fe} equals 54,055 mg kg^{-1} [26].

The EF and I_{geo} for Ti were calculated using the average concentrations of Ti (4.4 g kg^{-1}) and Fe (50 g kg^{-1}) in the earth's crust, instead of the QRVs for Cuban soils, because there is no reference value for Ti in Cuban soils:

$$EF_{Ti} = (C_{Ti}/C_{Fe})_{soil}/(C_{Ti}/C_{Fe})_{earth\ crust}$$

where C_{Ti} represents the Ti concentration and C_{Fe} represents the Fe concentration.

2.5. Statistical Analysis

Statistical analysis was performed by using the statistical software SPSS® 20.0 (New York, NY, USA). Analysis of variance (ANOVA) was used to assess the influence of district and soil use on the PTE concentrations. The Student's t-test was performed to compare pseudo-total and Mehlich-3 extractable metals in soils with and without gleyic properties and in surface and subsurface horizons. Principal component analysis (PCA) was applied to the pseudo-total data set for identifying associations among metals. Varimax was used as the rotation method in the PCA. The Kolmogorov–Smirnov (K–S) test was applied to check the normal distribution of data. Correlation coefficients were determined to complete and support the results obtained by factorial analysis.

3. Results

3.1. PTE Concentrations

The concentrations of pseudo-total and Mehlich-3 extractable and acid-oxalate extractable PTEs are presented in Table 1.

The Kolmogorov–Smirnov (K–S) test showed normal distributions for pseudo-total As, Mn, Ti, and V, Mehlich-3 Fe and Mn, and acid-oxalate Cu and Hg. The distribution was not normal for the remaining concentrations. Consequently, the Spearman's correlations were determined.

Pseudo-total and Mehlich-3 extractable metal concentrations were significantly correlated ($p < 0.01$) for all elements except Fe, As, and Hg. However, the proportions extracted by Mehlich-3 reagent varied widely among elements. The metal extracted by the Mehlich-3 reagent was a negligible fraction of the pseudo-total metal in the cases of Ti (0.11%) and V (0.07%). This fraction was also very low for Fe (0.66%), As (1.02%), Hg (1.29%), and Cr (1.33%). On the contrary, high fractions of Cd and Pb (30%), Mn (22%), Zn and Cu (18%) were extracted by Mehlich-3 reagent. Intermediate values were found for Co (13%) and Ni (6%).

Acid-oxalate concentrations correlated to pseudo-total concentrations ($p < 0.01$) for Co, Cr, Fe, Mn, and Pb, and to Mehlich-3 concentrations ($p < 0.05$) for Co, Cr, Cu, Mn, Ni, and Pb.

Table 1. Concentrations of potentially toxic elements (PTEs) for the 35 soils studied: average ± standard deviation, along with world soil average and maximum allowable concentrations (MAC) for agricultural soils according to Kabata-Pendias [27] (see also Table S1).

	Pseudo-Total, mg kg^{-1}	World Average, mg kg^{-1}	MAC, mg kg^{-1}	Mehlich 3, mg kg^{-1}	Acid Oxalate, mg kg^{-1}	100 × Mehlich 3/Pseudo-Total
As	8.11 ± 3.09	6.83	15–20	0.074 ± 0.054	0.098 ± 0.082	1.02 ± 0.66
Cd	0.62 ± 0.56	0.41	1–5	0.187 ± 0.155	n.d.	30.4 ± 10.4
Co	11.7 ± 7.8	11.3	20–50	1.56 ± 1.37	1.35 ± 1.24	12.5 ± 7.5
Cr	82.9 ± 119.3	59.5	50–200	0.40 ± 0.25	2.27 ± 2.46	1.33 ± 2.70
Cu	73.5 ± 37.5	38.9	60–150	14.1 ± 11.0	1.78 ± 0.80	18.2 ± 6.7
Fe	25,500 ± 9960	35,000	—	124 ± 38	1490 ± 1110	0.66 ± 0.54
Mn	578 ± 235	488	—	131 ± 82	72.8 ± 57.5	21.9 ± 10.9
Ni	72.1 ± 131.0	29	20–60	2.95 ± 3.22	2.11 ± 3.48	6.34 ± 3.25
Pb	73.5 ± 79.4	27	20–300	25.0 ± 33.3	0.69 ± 0.57	30.4 ± 11.8
Ti	907 ± 393	7,038	—	0.93 ± 0.41	2.60 ± 2.37	0.11 ± 0.07
V	619 ± 253	129	150	0.41 ± 0.23	6.03 ± 4.36	0.07 ± 0.07
Zn	126 ± 88	70	100–300	27.2 ± 33.9	16.7 ± 9.3	18.3 ± 9.7
Hg	0.51 ± 0.57	0.07	0.5–5	0.0017 ± 0.0011	0.041 ± 0.029	1.29 ± 1.91

The percentage of metal extracted by acid oxalate, referred to the pseudo-total metal, was very high in the case of Hg, with an average value of 32% and a maximum value of 96%. The extractions were high for Co, Mn, and Zn (with average values of 10, 11, and 12%, respectively, referring to the pseudo-total concentration). For the other metals, the extraction percentages were low, with average values of less than 3%.

3.2. Analysis of Variance

The analysis of variance showed that the district significantly affected the pseudo-total concentrations of As, Co, Cr, Ni, Ti, V, Hg ($p < 0.001$), and Mn ($p < 0.05$) (Table 2). On the contrary, the pseudo-total concentrations of Cd, Cu, Fe, Pb, and Zn were not significantly different among districts.

The land use did not affect significantly the pseudo-total concentrations of the elements studied.

The district of Regla had the highest concentrations of pseudo-total Co, Cr, and Ni (Figure 2). The highest concentrations of As were found in Playa and Plaza de la Revolución, while the highest concentrations of Hg were found in Playa, 10 de Octubre, and Habana Vieja. Playa had the highest concentrations of Mn, and the highest concentrations of Ti and V were found in San Miguel del Padrón and Regla.

Table 2. Main effects of district, soil use, and interaction district x soil use on pseudo-total PTE concentrations (significance level).

	District	Significance Soil Use	District x Soil Use
As	**0.00**	0.09	0.52
Cd	0.49	0.40	0.25
Co	**0.00**	0.49	0.71
Cr	**0.00**	0.53	**0.03**
Cu	0.54	0.14	0.63
Fe	0.16	0.05	0.31
Mn	**0.05**	0.27	0.53
Ni	**0.00**	0.52	0.18
Pb	0.66	0.23	0.10
Ti	**0.00**	0.16	0.99
V	**0.00**	0.15	0.95
Zn	0.12	0.88	0.88
Hg	**0.00**	0.13	**0.01**

Bold numbers indicate significance level ≤0.05.

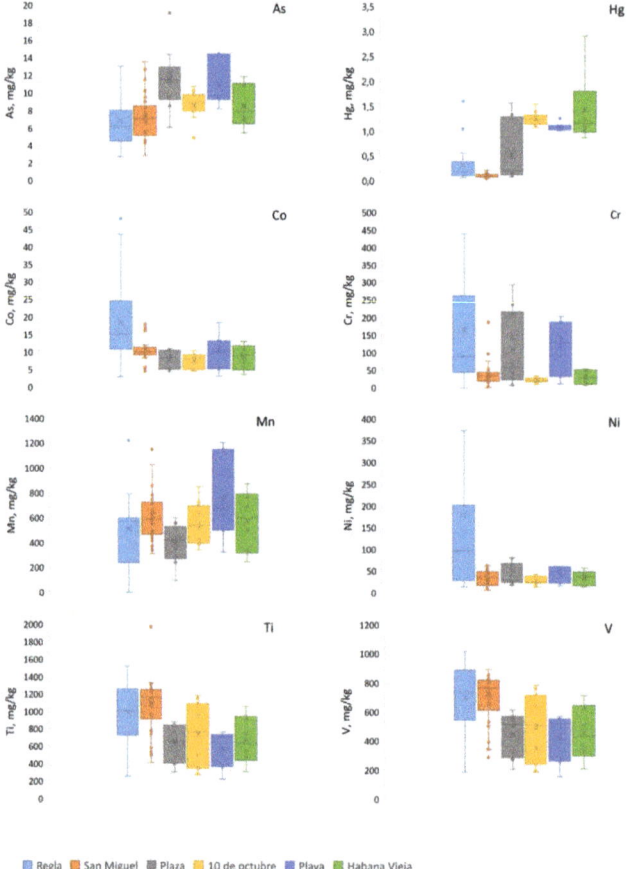

Figure 2. Distribution of pseudo-total As, Hg Co, Cr, Mn, Ni, Ti, and V in the various districts. The limits of the boxes are percentiles 25 and 75. The horizontal lines inside the boxes are the medians. The whiskers indicate the 0th and 100th percentiles, excluding outliers.

3.3. Principal Component Analysis

The results of PCA for pseudo-total metal concentrations in soils are shown in Tables 3 and 4. Four principal components (PC) with eigenvalues higher than 1 (before and after rotation) were extracted. The total variance explained by these four components is nearly 78%. In the rotated matrix, Ti, V, Fe (with positive loadings) and As and Hg (with negative loadings) are associated in a first component (PC1), which explains 24.8% of the variance. A second component (PC2), which combines Ni, Cr, and Co (all with positive loadings), and explaining 22.0% of variance, was identified. The third component (PC3) combines Pb, Zn, Cd, and Cu, accounting for 19.0% of variance. The fourth component (PC4) includes only Mn and accounts for 11.8% of variance.

3.4. Correlation Matrix

Table 5 presents the correlation matrix among pseudo-total metals. In general, the correlations were in accordance with the results obtained by PCA.

Interestingly, all elements except Cd, Pb, Zn, and Hg were significantly ($p < 0.01$) correlated with Fe, a mostly lithogenic element. The correlation was negative for As and positive for the other elements.

Table 3. Total explained variance.

Component	Initial Eigenvalues			Rotation Sum of Squared Loadings		
	Total	% of Variance	% Cumulative	Total	% of Variance	% Cumulative
1	4.211	32.39	32.39	3.225	24.81	24.81
2	2.799	21.53	53.92	2.866	22.05	46.85
3	2.041	15.70	69.62	2.476	19.05	65.90
4	1.055	8.12	77.74	1.539	11.84	77.74
5	0.792	6.09	83.83			
6	0.627	4.83	88.66			
7	0.518	3.98	92.64			
8	0.357	2.75	95.38			
9	0.302	2.33	97.71			
10	0.209	1.61	99.31			
11	0.074	0.57	99.89			
12	0.015	0.11	100.0			
13	0.000	0.00	100.0			

Table 4. Matrix of principal components analysis (significant loading factors are highlighted in bold).

	Component Matrix				Rotated Component Matrix				
Element	PC1	PC2	PC3	PC4	Element	PC1	PC2	PC3	PC4
Fe	**0.828**	0.092	0.201	0.157	Ti	**0.950**	−0.094	0.027	0.124
Co	**0.818**	0.299	−0.414	0.050	V	**0.947**	0.056	0.048	0.137
V	**0.762**	−0.393	0.393	−0.180	As	**−0.707**	−0.250	0.222	−0.257
As	**−0.724**	0.0392	−0.011	−0.009	Fe	**0.577**	0.354	0.250	0.488
Ti	**0.669**	−0.474	0.0480	−0.158	Hg	**−0.575**	−0.086	0.245	0.193
Cr	**0.592**	0.474	−0.569	−0.181	Ni	0.034	**0.976**	0.022	−0.018
Zn	0.067	**0.761**	0.380	0.156	Cr	0.010	**0.959**	0.104	0.040
Cd	0.008	**0.709**	0.387	−0.084	Co	0.250	**0.867**	0.047	0.340
Pb	−0.054	**0.605**	0.475	−0.422	Pb	−0.013	−0.012	**0.821**	−0.314
Hg	−0.359	**0.447**	−0.008	0.326	Zn	−0.216	0.068	**0.795**	0.264
Ni	0.0588	0.403	**−0.627**	−0.231	Cd	−0.158	0.057	**0.795**	0.017
Cu	0.446	0.355	**0.448**	−0.000	Cu	0.325	0.108	**0.591**	0.242
Mn	0.522	0.077	0.085	**0.754**	Mn	0.140	0.125	0.034	**0.905**

Al metals associated in PC1 with positive loadings (Ti, V, and Fe) correlated positively with each other (p < 0.01). As and Hg correlated positively with each other (p < 0.05) and negatively with Fe, Ti, and V (p < 0.01), except Hg with Fe. Remarkably, the correlation between Ti and V was extremely high (r = 0.980, p = 0.0000).

The three elements combined in PC2 (Ni, Cr, and Co) correlated positively to each other (p < 0.01). Furthermore, the four elements combined in PC3 (Pb, Zn, Cd, and Cu) correlated positively to each other (p < 0.01).

Table 5. Spearman's correlation coefficient matrix between pseudo-total metal concentrations in the soils studied.

	As	Cd	Co	Cr	Cu	Fe	Mn	Ni	Pb	Ti	V	Zn
Cd	0.27 *											
Co	−0.59 **	0.14										
Cr	−0.04	0.58 **	0.57 **									
Cu	−0.18	0.47 **	0.56 **	0.44 **								
Fe	−0.49 **	0.20	0.70 **	0.49 **	0.55 **							
Mn	−0.40 **	0.10	0.63 **	0.12	0.32 **	0.40 **						
Ni	−0.07	0.49 **	0.56 **	0.90 **	0.46 **	0.39 **	0.20					
Pb	0.17	0.58 **	0.08	0.29 **	0.49 **	0.03	−0.06	0.33 **				
Ti	−0.70 **	−0.21 *	0.48 **	−0.06	0.39 **	0.53 **	0.24 *	−0.07	0.04			
V	−0.72 **	−0.18	0.54 **	0.04	0.42 **	0.59 **	0.26 *	0.04	0.05	0.98 **		
Zn	0.18	0.79 **	0.18	0.45 **	0.57 **	0.21	0.05	0.44 **	0.77 **	−0.08	−0.05	
Hg	0.26 *	0.34 **	−0.06	0.14	0.14	−0.04	−0.06	0.20	0.42 **	−0.35 **	−0.31 **	0.53 **

* significant at p < 0.05 level; ** significant at p < 0.01 level.

3.5. Influence of Soil Depth and Gleyic Properties

The concentrations of pseudo-total and bioavailable Cd and Zn, pseudo-total Cr and Ni, and bioavailable Cu were higher in surface horizons (Table 6).

The concentrations of pseudo-total Co, Cr, Fe, and Ni were higher in soils with gleyic properties than in soils without gleyic properties (Table 7). Furthermore, the concentrations of bioavailable Co, Ni, and Ti were higher in soils with gleyic properties than in soils without gleyic properties.

Table 6. Mean values and standard deviations of the concentrations (mg kg^{-1}) of pseudo-total and Mehlich-3 extractable PTEs from surface (n = 35) and subsurface (n = 56) horizons of the urban soils studied and significance of equality of means (T-test). Significant differences (p ≤ 0.05) highlighted in bold.

	Pseudo-Total Elements, mg kg^{-1}			Mehlich-3 Extractable Elements, mg kg^{-1}		
	Surface	Subsurface	p	Surface	Subsurface	p
As	7.74 ± 2.34	8.35 ± 3.48	0.37	0.086 ± 0.047	0.066 ± 0.056	0.07
Cd	**0.788 ± 0.681**	**0.514 ± 0.433**	**0.04**	**0.240 ± 0.168**	**0.153 ± 0.137**	**0.01**
Co	13.4 ± 9.9	10.7 ± 6.0	0.16	1.71 ± 1.49	1.47 ± 1.30	0.43
Cr	**116 ± 160**	**61.8 ± 78.5**	**0.03**	0.433 ± 0.258	0.372 ± 0.247	0.27
Cu	80.6 ± 37.4	69.1 ± 37.2	0.16	**17.2 ± 13.2**	**12.2 ± 9.0**	**0.05**
Fe	27,300 ± 9020	24,400 ± 10,400	0.16	122 ± 31	125 ± 42	0.68
Mn	588 ± 217	573 ± 247	0.76	171 ± 200	125 ± 85	0.20
Ni	**103 ± 190**	**53.1 ± 69.7**	**0.05**	3.25 ± 4.09	2.77 ± 2.57	0.53
Pb	84.7 ± 60.9	66.5 ± 88.9	0.25	25.8 ± 18.0	24.5 ± 40.1	0.83
Ti	847 ± 307	944 ± 437	0.25	0.997 ± 0.458	0.879 ± 0.378	0.25
V	593 ± 194	635 ± 283	0.44	0.401 ± 0.264	0. 417 ± 0.201	0.76
Zn	**163 ± 108**	**103 ± 65**	**0.01**	**41.1 ± 46.9**	**18.6 ± 18.0**	**0.00**
Hg	0.528 ± 0.544	0.491 ± 0.591	0.76	0.0016 ± 0.0008	0.0017 ± 0.0012	0.39

Table 7. Mean values and standard deviations of the concentrations (mg kg^{-1}) of pseudo-total and Mehlich-3 extractable PTEs of the urban soils studied, without gleyic properties (n = 62) and with gleyic properties (n = 29), and significance of equality of means (T-test). Significant differences (p ≤ 0.05) highlighted in bold.

	Pseudo-Total Elements. mg kg^{-1}			Mehlich-3 Extractable Elements. mg kg^{-1}		
	Nongleyic	Gleyic	p	Nongleyic	Gleyic	p
As	7.96 ± 3.31	8.46 ± 2.56	0.44	0.079 ± 0.056	0.062 ± 0.046	0.13
Cd	0.620 ± 0.532	0.618 ± 0.614	0.98	0.184 ± 0.155	0.193 ± 0.156	0.79
Co	**13.2 ± 8.9**	**8.43 ± 2.29**	**0.01**	**1.74 ± 1.47**	**1.17 ± 1.04**	**0.04**
Cr	**101 ± 134**	**43.3 ± 65.1**	**0.03**	0.420 ± 0.247	0.340 ± 0.259	0.16
Cu	75.6 ± 34.6	69.0 ± 43.6	0.49	14.9 ± 11.3	12.4 ± 10.4	0.33
Fe	**27,100 ± 10,100**	**22,000 ± 8850**	**0.02**	124 ±40	124 ± 35	0.99
Mn	597 ± 257	537 ± 176	0.20	154 ± 162	119 ± 82	0.18
Ni	**89.9 ±154.1**	**32.2 ± 18.4**	**0.05**	**3.45 ± 3.75**	**1.83 ± 0.72**	**0.03**
Pb	66.2 ± 68.2	89.8 ± 99.8	0.19	23.4 ± 33.5	28.5 ± 33.1	0.51
Ti	941 ± 411	831 ± 344	0.19	**1.00 ± 0.36**	**0.702 ± 0.482**	**0.01**
V	646 ± 263	559 ± 220	0.11	0.421 ± 0.242	0.378 ±0.175	0.41
Zn	129 ± 94	121 ± 75	0.68	28.5 ± 38.1	24.5 ± 22.3	0.53
Hg	0.463 ± 0.562	0.601 ± 0.586	0.30	0.0018 ± 0.0013	0.0015 ± 0.0004	0.24

4. Discussion

The pseudo-total PTE concentrations (Table 1) were generally higher than the average values for the world's soils: 58% of the samples presented pseudo-total As concentrations higher than the average concentrations for world's soils reported by Kabata-Pendias [27]; this percentage was also 58% for Cd, 33% for Co, 31% for Cr, 87% for Cu, 65% for Mn, 58% for Ni, 70% for Pb, 71% for Zn, and 84% for Hg. Furthermore, the average V concentration for world's soils was exceeded by all samples, while the average Ti concentration for world's soils was never exceeded. However, Ti and V pseudo-total concentrations were highly correlated.

Conversely, the measured pseudo-total metal concentrations were similar to those published by other authors for urban soils [6,11–15,28–30], except for vanadium, which presented very high concentrations in the soils studied.

The pseudo-total concentrations for most PTEs did not exceed the maximum concentrations allowed in agricultural soils. The maximum allowable concentrations (MAC) in agricultural soils reported by Kabata-Pendias [27] for Cr and Ni were exceeded by 12% and 22% of samples, respectively. As discussed below, these elements have a mainly lithogenic character, associated with ultramafic rocks. The MAC in agricultural soils for Cu, Pb, and Zn were exceeded by 4% (4 samples), 1% (one sample), and 3% (3 samples) of samples, respectively. The V concentrations were very high, always exceeding the MAC for agricultural soils. This element, as discussed below, can have a lithogenic character, although anthropogenic sources might contribute to increase its concentrations.

The measured pseudo-total Co, Cu, Ni, and Pb are in accordance with those reported by Diaz-Rizo et al. [31] for surface soils (0–10 cm) of Havana, measured by X-ray fluorescence. The Zn concentrations were mostly lower in the present study, rather closer to the values for nonurbanized areas in [31].

Cuban soils have naturally high concentrations of some heavy metals, exceeding usual quality standards [26]. This is true especially for Ni, Cr, Cu, and Co, abundant in ultramafic rocks, which are common in the island. Consequently, Rodriguez Alfaro et al. [26] established quality reference values (QRVs) for PTEs in Cuban soils, based in the analysis of representative natural soils (Table 8).

The pseudo-total concentrations of all the elements determined by Rodriguez Alfaro et al. [26] in Cuban natural soils correlated significantly with the concentration of the lithogenic element Fe. Similarly, in the present study this happens to be true for all the elements studied except Cd, Pb, Zn, and Hg, which is a clear indication of the human origin of these elements.

Table 8. Quality reference values (QRVs) established by Rodriguez Alfaro et al. (2015) for Cuban soils, enrichment factors (EF) and geoaccumulation indices (I_{geo}) calculated for the urban soils studied by comparison with these QRVs (median and interval in parentheses).

	QRV, mg kg^{-1}	EF	I_{geo}
As	19	0.75 (0.19 to 5.17)	−1.87 (−3.40 to −0.58)
Cd	0.6	1.84 (0.89 to 12.12)	−0.94 (−3.91 to 2.00)
Co	25	0.90 (0.42 to 3.20)	−1.89 (−3.67 to 0.36)
Cr	153	0.67 (0.01 to 7.08)	−2.59 (−8.06 to 1.84)
Cu	83	1.67 (0.94 to 6.78)	−0.86 (−3.30 to 0.88)
Mn	1947	0.65 (0.08 to 3.24)	−2.35 (−4.94 to −1.25)
Ni	170	0.54 (0.08 to 7.26)	−2.74 (−5.08 to 1.87)
Pb	50	2.19 (0.15 to 50.1)	−0.61 (−4.85 to 2.49)
Ti	—	0.45 (0.16 to 0.87) *	−2.80 (−4.88 to −1.55) *
V	137	10.5 (4.20 to 19.5)	1.65 (−0.40 to 2.82)
Zn	86	2.65 (0.89 to 12.2)	−0.31 (−2.70 to 2.37)
Hg	0.1	2.70 (0.90 to 176)	0.15 (−1.62 to 4.28)

* In the case of Ti, the average concentrations of Ti and Fe in the earth's crust were used as a reference.

These QRVs for Cuban soils were used as background concentrations to calculate the EF and Igeo for PTEs in urban soils.

The EFs for Ti were always lower than 1, indicating no enrichment, while the calculated I_{geo} were always negative, indicating the absence of pollution. Titanium is a lithogenic element, and a common constituent of several rocks [27].

Except for Ti, the lowest enrichment factors (Table 8) were calculated for As, Co, Cr, Mn, and Ni. These EFs were lower than 2 (minimum enrichment) for most samples, although there are a few samples with values between 2 and 5 (moderate enrichment) for Co, Cr, Mn, and Ni and 19% of samples (in all districts) with EFs between 2 and 5 for As. Co, Cr, Mn, and Ni are lithogenic elements, Co, Cr, and Ni being associated with ultramafic rocks and Mn with both acid and basic rocks [27]. However, soil enrichment can occur through anthropogenic emissions. The application to soils of composts and organic amendments can contribute with several PTEs, such as As, Co, Cr, Cu, Mn, Ni, Pb, and Zn [32].

According to the EF values, there was no or minimum enrichment in Co and Mn for most samples, with a few outliers showing moderate enrichment (Figure 3). There was minimum enrichment in As, Cd, Cr, Cu, and Ni for most samples, but outliers showed moderate (EF between 2 and 5) to significant (EF between 5 and 20) enrichment. These elements can be of lithological origin, but the presence of outliers suggest the contribution of anthropogenic sources in some samples. These possible anthropogenic sources include the traffic of vehicles [13], metal smelting, waste disposal and incineration, burning of fossil fuels [33,34], and the application of organic amendments to soils [35].

The enrichment factors were high or very high for Pb, V, Zn, and Hg (Table 8, Figure 3), which points to a human origin of these elements. This origin is unquestionable for Pb and Hg. The past use of leaded gasoline, Pb-containing paints, metal smelting, and fuel combustion, among other anthropogenic activities, can be sources of lead [27,36]. Mercury can originate from combustion of coal and oil, smelting industry, or waste incineration [27,36]. Zn can be lithogenic, but it can also be brought to the soil with organic amendments [32,35] or as a result of various human activities, such as metal smelting or traffic of vehicles. V can have a lithogenic character, mainly associated to ultramafic rocks, but its enrichment may be linked to anthropogenic activities, such as combustion of fuel oil, metal smelting, or oil refining [27,36]. There was significant enrichment in V, while for Pb, Zn, and Hg the median values denote moderate enrichment, but outliers reached significant enrichment in Zn and extremely high enrichment in Pb and Hg.

The values of the geoaccumulation index (I_{geo}) were always negative, indicating the absence of pollution, for As, Mn, and Ti, even in cases of moderate enrichment in As and Mn. For Co and Cu, the I_{geo} were negative for all but one (Co) or six (Cu) samples, which had I_{geo} values between 0 and 1 (unpolluted to moderately polluted). In the cases of Cd, Cr, and Ni, the I_{geo} values ranged from negative to 2.00, 1.84, and 1.87, respectively, indicating situations between practically unpolluted and moderately polluted. The I_{geo} values reached 2.49, 2.82, and 2.37 for Pb, V, and Zn, respectively, indicating some situations of moderate to strong pollution ($2 < I_{geo} < 3$). The worse situation is that of Hg, with 13 samples (14%) with I_{geo} values between 3 and 4 (strongly polluted) and 1 sample with $I_{geo} = 4.28$ (strongly to extremely polluted). In general, the interpretation of I_{geo} values is in accordance with that of EF values.

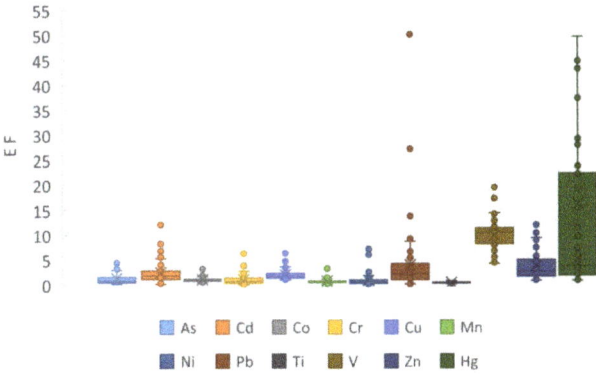

Figure 3. Enrichment factors for the studied elements relative to Cuban natural soils.

The first principal component in the PCA, accounting for 25% of the variance in the rotated matrix (Tables 3 and 4), groups the elements Ti, V, Fe, As, and Hg. Ti, V, and Fe, correlating positively to each other (Table 5), must be mostly lithogenic. Conversely, As and Hg, correlating negatively with Ti, V, and Fe, seem to be associated with soil organic fraction and have an anthropogenic origin. Both elements can originate from metal processing industry, combustion of fuel, waste incineration, and application of organic amendments to soil [27,36].

The vanadium enrichment could be attributed to smelting industries and to generators located in all the Havana districts, since V can be present as an impurity in fuels. The combustion of oil and its derivatives is considered the main source of vanadium contamination. V concentrations in crudes from Venezuela are particularly high, ranging from 282 to 1180 mg kg^{-1} [37].

As referred to before, the titanium and vanadium concentrations showed a very high correlation (Table 5), which points to a common origin. Vanadium appears as an impurity in titanium minerals, such as ilmenite and rutile. Both elements are present in pyroxenes and amphiboles, being, therefore, relatively more abundant in basic and ultrabasic rocks [27]. Furthermore, both elements are used in the steel industry. Therefore, its origin seems to be mainly lithological, although there seems to be an anthropogenic enrichment of vanadium.

The second principal component includes the elements Ni, Cr, and Co and accounts for 22% of the variance in the rotated matrix (Tables 3 and 4). These elements, highly correlated to each other (Table 5), are abundant in ultramafic rocks. They are therefore considered as lithogenic. Anthropogenic activities (fuel combustion, smelting industry, and soil organic amendments) may contribute to increasing the concentrations of these elements in some samples. Accordingly, Díaz-Rizo et al. [38] report high bivariate correlations among Co, Cr, and Ni in soils of Las Tunas City (NE Cuba), which they interpret as a result of a common natural (nonanthropogenic) origin.

The third principal component, which explains 19% of the variance, includes Pb, Zn, Cd, and Cu (Tables 3 and 4). Cd, Pb, and Zn did not significantly correlate with Fe, which indicates a nonlithogenic origin. The association in the same PC points to an anthropogenic origin of the four elements. The higher concentrations of Zn and Cd (both pseudo-total and bioavailable) and bioavailable Cu in surface horizons (Table 6) are in accordance with this anthropogenic origin. The surface soil is usually more likely affected by anthropogenic sources than the subsurface horizons, closer to the parent material. However, urban soils in general, and the soils studied here in particular, are often allochthonous and, therefore, the enrichment of the surface soil in a certain element does not necessarily indicate a nonlithogenic character, since different horizons may have different origins. Our results are in accordance with those reported by Díaz-Rizo et al. [31], which, based on the bivariate correlations among Cu, Zn, and Pb, suggest a common origin for these three elements in Havana urban soils. However, by comparing soils from urban and nonurbanized areas, these authors conclude that Pb and Zn have a human origin, while Cu, the same as Co and Ni, may have a lithogenic character. Furthermore, for Las Tunas soils, Díaz-Rizo et al. [38] report a high correlation between Pb and Zn, which are considered anthropogenic elements. Because Cu is an abundant element in ultramafic rocks, its presence in Havana soils may have in part a geological origin, but the PCA points to an important anthropogenic contribution. The absence of significant correlations between Cu and any of Co, Cr, Ni, Pb, and Zn and their different spatial distribution in the aforementioned study from Las Tunas [38] would be in accordance with the double lithogenic and anthropogenic character of Cu.

Similarly to our results, in a study of urban soils of Hong Kong [16], Cd, Cu, Ni, Pb, and Zn were associated in the same principal component and considered anthropogenic. Massas et al. [11] and Gasparatos et al. [12] inferred from their data for urban soils of Greece a common anthropogenic origin for Zn, Cu and Pb. Also in Vigo (Spain), Pb, Cu and Zn were identified as anthropogenic [13]. Sun et al. [6] reported an association among the anthropogenic Cd, Cu, Pb and Zn in urban soils of Shenyang, China. In topsoils of urban parks in Beijing, China [33], Pb, Zn, and Cu were strongly correlated and were considered of anthropogenic origin, mainly related to vehicular traffic. In summary, profuse research shows that Cd, Cu, Pb, and Zn are ubiquitous in urban soils and arise from human activity.

The last component (PC4) explains 11% of the variance and includes only Mn (Tables 3 and 4). This is usually a lithogenic element. However, its isolation in a unique component and its high concentration in some samples (up to 1230 mg kg^{-1}) could indicate a partially human origin. The major anthropogenic sources of Mn are municipal wastewaters, sewage sludge, coal combustion, and metal smelting processes [27,36].

Our results are broadly consistent with those of Manta et al. [14], which for urban soils of Palermo (Italy) conclude that Pb, Zn, Cu, Sb, and Hg can be considered tracers of anthropogenic pollution, whereas Mn, Ni, Co, Cr, V, and Cd were interpreted to be mainly inherited from parent materials.

Ti and V, having a common origin, showed the same spatial distribution (Figure 2). Similarly, Ni, Cr, and Co presented an analogous spatial distribution. On the contrary, the anthropogenic As and Hg showed different spatial distribution patterns. The spatial distribution of Mn was different from any other element. Cd, Cu, Fe, Pb, and Zn were not affected by the district; this is interpreted as a result of lithological origin for Fe and ubiquitous contamination with Cd, Cu, Pb, and Zn.

The high levels of Ti and V can be linked to the oil refinery Ñico López and the electric power plant Antonio Maceo in Regla and to the smelting industry FUNALCO in San Miguel del Padrón, as well as to various small sources like electricity generating facilities. The highest Hg concentrations in Habana Vieja, the historical centre of the city, are in accordance with reports of a relationship between the accumulation of Hg and the duration of urbanization [33]. High concentrations of Co, Cr, and Ni (Figure 2) can be related to lithological factors (ultramafic rocks). They could also be associated with the oil refinery Ñico López, in Regla, with the traffic of vehicles and with the application of organic amendments to soils. All the enriched elements can originate from local application of different organic amendments.

The high concentrations of PTEs, mainly Cr, Ni, and V, in some soils pose an environmental risk, particularly in soils for agricultural use. However, it should be noted that these elements are expected to be scarcely soluble in moderately alkaline soils, such as those in this study. Moreover, to the extent that they are lithogenic elements, they must be part of primary minerals and be released only upon weathering of these minerals. Chromium is solubilized above pH 8 as CrO_4^{2-} [27]. Therefore, an increase of soil pH would be risky, taking into account that Cr (VI) is carcinogenic to humans. The high content of organic matter in soil favors the reduction of Cr (VI) to Cr (III), less mobile and less toxic to human. Vanadium is found in neutral and alkaline soils in anionic forms, which are adsorbed by iron oxides or clay minerals through Fe cation bridges [27]. Nickel is rather mobile in horizons rich in organic matter, where it is solubilized in the form of chelates, although the complexation by organic ligands reduces the toxicity of Ni for soil organisms.

Pb, Zn, and Hg, having high enrichment factors and geoaccumulation indices, do not exceed the upper limits of the maximum allowable concentrations for agricultural soils (Table 1).

To evaluate the risk associated to the presence of PTEs in soils, it is necessary not only to know the total metal concentrations, but also the concentrations of mobile or bioavailable forms. The soil pH, the presence of carbonates, the organic matter content, the redox potential, and the presence of potential adsorbents, such as iron, aluminum, and manganese oxyhydroxides, are, among others, factors that influence the retention/mobility of PTEs.

The Mehlich 3 reagent extracts bioavailable PTEs [24,39,40]. The low proportions of Mehlich-3 extractable Ti (0.10%), V (0.07%), Fe (0.66%), As (1.02%), Hg (1.44%), and Cr (1.60%) indicate a low mobility and a low risk of exportation of these elements to food chain or water bodies. This is particularly relevant for vanadium and mercury: even though the pseudo-total V concentrations were very high, the bioavailable V concentrations (Table 1) were very low; even though the EFs of Hg were very high, the bioavailable Hg concentrations were also very low, suggesting that there is no risk of V or Hg being exported to vegetation or water bodies. This risk is also low for Cr, whose pseudo-total concentrations were high in certain samples. According to Baken et al. [41], vanadium is strongly adsorbed onto poorly crystalline oxyhydroxides. This is in accordance with the fact that V extracted by acid oxalate (associated with poorly crystalline oxyhydroxides) is considerably higher than V extracted by Mehlich-3 reagent (bioavailable, Table 1). For soils derived from fluvioglacial sands in Poland, Jeske and Gworek [42] reported low mobilities of Cr, Ni, and V, the latter being the least mobile among them. According to Larsson et al. [43], V toxicity to plants and soil microorganisms is controlled by the vanadium sorption capacity of soils and, therefore, V concentration in the soil solution is a better predictor of V toxicity than total V in soil.

On the contrary, high proportions of metals in available form (observed for Mn, Cd, Pb, Zn, and Cu) indicate a recent enrichment, so that there has been no time for these metals to be immobilized in organic or inorganic forms [11,15]. In accordance with our results, Massas et al. [11] reported very low availability ratios (available metal/pseudo-total metal) for Cr and high availability ratios for Pb, Cu, and Zn in urban soils in Greece. Similarly, Pb was the most mobile and Cr the least mobile metal, among Pb, Zn, Cu, Ni, and Cr, in urban soils of the same area [12].

For Ni, whose pseudo-total concentrations exceeded the MAC for agricultural soils in 22% of samples, some values of the concentrations in bioavailable form are still high. The critical toxicity limit according to Monterroso et al. [39] (7 mg kg^{-1} at pH 8) was exceeded by three soils of the district Regla, one of them being an agricultural soil. However, the determined Mehlich-3 extractable concentrations were all within the range reported by Caridad-Cancela et al. [44] for natural soils of Galicia, NW Spain (0.22–42.65 mg kg^{-1}).

The toxicity limit for bioavailable Cd (0.25 mg kg^{-1}) was exceeded in some horizon of five agricultural soils and three park or vacant soils. Nevertheless, it is worth mentioning that the determined concentrations were very similar to those reported by Caridad-Cancela et al. [44] for natural soils of Galicia, NW Spain (<0.01–0.41 mg kg^{-1}). The toxicity limits for Mn (140 mg kg^{-1}) and Pb (12 mg kg^{-1}) were often (39% and 56% of samples, respectively) exceeded by soils of various uses. One sample had a concentration of bioavailable Cu above the toxicity limit (60 mg kg^{-1}), and two samples had a concentration of bioavailable Zn above the toxicity limit (100 mg kg^{-1}).

According to Rodriguez Alfaro et al. [45], high concentrations of PTEs can be present in Cuban agricultural soils as a result of the application of municipal-solid-waste compost, which contain elevated concentrations of Cd, Pb, Hg, As, Se, and Ni. According to these authors, as a result of their study, the use of municipal-solid-waste composts for food production has been forbidden by Cuban authorities. Upon discontinuation of the application of contaminated municipal-solid-waste compost, the PTEs will expectedly evolve into immobile forms.

The extraction with acid oxalate allows estimation of the elements occluded in noncrystalline iron oxides, which can be made available in the medium to long term [46], particularly if a change in soil conditions (e.g., pH and redox potential) favors the dissolution of oxides. This would be of particular concern for Hg, although the pseudo-total concentrations for this metal were always well below the reported upper limit of the MAC for agricultural soils.

The existence of gleyic properties had a significant influence on the concentrations of PTEs, both pseudo-total and available. The pseudo-total Fe, Co, Cr, and Ni concentrations were significantly ($p < 0.05$) lower in the soils with gleyic properties. Mobility and subsequent loss of iron are well-known facts in soils with reducing conditions. Co, Cr, and Ni can coprecipitate with iron or be adsorbed or occluded in iron oxyhydroxides, so that they are mobilized upon solubilization. It may also happen that these elements are mobilized in the form of complexes with soluble organic matter, more abundant under reducing conditions. Similarly, Shaheen et al. [47] reported a release of Cd, Cu, Co, Fe, Mn, Ni, and Zn under flood-dry cycles in a contaminated floodplain soil, related to variations of redox potential, pH, and dissolved organic carbon. In laboratory experiments, waterlogging was observed to affect the mobility of heavy metals (Cu, Pb, and Zn) retained or occluded in Fe or Mn compounds in urban soils [34,48].

The soils with gleyic properties showed lower concentrations ($p < 0.05$) of Mehlich-3 extractable Co, Ni, and Ti. In the cases of Co and Ni, this behavior was the same as that observed in pseudo-total concentrations. For these elements, pseudo-total and Mehlich-3 extractable concentrations were highly correlated. In the case of Ti, there were no significant differences in pseudo-total concentrations between gleyic and nongleyic soils. The extractable Ti/pseudo-total Ti ratio was significantly lower in soils with gleyic properties ($p < 0.01$). This appears to be a loss of Ti solubility under reducing conditions, which could be due to Ti precipitation as sulfide. This precipitation as sulfide may also occur for Co and Ni; in these cases, the precipitation, added to the decrease in total contents of these elements, could intensify the decline of available Co and Ni. Therefore, a change in the redox conditions of soils (for example, a variation of the water table) can result in mobilization or immobilization of some PTEs, increasing or decreasing the risk for these metals being exported to the food chain or water bodies. According to Ajmone-Marsan et al. [34], waterlogging of urban soils contaminated with PTEs may pose a serious environmental risk, particularly in the context of climate change.

5. Conclusions

The study shows that Fe, Ti, V, Ni, Cr, and Co were mainly of lithological origin, although a contribution of anthropogenic sources was not negligible for V, Ni, and Cr. As, Hg, Pb, Zn, Cd, and Cu arose mostly from anthropogenic activities, being related to industrial activities, fuel combustion, and application of organic amendments to soil. Mn seems to have a combined origin (lithogenic and anthropogenic).

Ti and V, having a common origin, showed the same distribution among the studied districts. Furthermore, Ni, Cr, and Co, associated with ultramafic rocks, showed a similar spatial distribution.

The pseudo-total concentrations of Cr and Ni were above the maximum allowable concentrations (MAC) for agricultural soils in 12% and 22% of samples, respectively, while the concentrations of Cu, Pb, and Zn rarely exceeded the MAC values. The pseudo-total V concentration always exceeded this MAC. For the other elements, the concentrations were below the MAC.

The enrichment factors (EFs) showed no anthropogenic enrichment for Ti. There was minimum enrichment of most samples in Co, Mn, As, Cd, Cr, Cu, and Ni, but outliers reached moderate enrichment in Co, Mn, and As, and significant enrichment in Cd, Cr, Cu, and Ni. There was significant enrichment in V, while for Pb, Zn, and Hg the median values denote moderate enrichment, but outliers reached significant enrichment in Zn and extremely high enrichment in Pb and Hg.

The concentrations of bioavailable forms of metals (including some usually considered as lithogenic) were quite often above acceptable thresholds, suggesting a recent enrichment from anthropogenic sources. The toxicity limits for bioavailable Cd, Ni, Mn, and Pb were exceeded by 14%, 10%, 39%, and 56% of samples, respectively.

The results of this study may be valuable for the authorities to issue guidelines for better management of the urban agriculture activities underlining the most toxic elements as well as their safety limits.

Supplementary Materials: The following are available online at http://www.mdpi.com/2076-3298/7/6/43/s1, Table S1: Concentrations of potentially toxic elements in urban soils of Havana. Raw data.

Author Contributions: Conception and design of the work: J.M.M.-A. and R.O.-G. Acquisition of data: J.M.M.-A. Analysis and interpretation of data: J.M.M.-A., R.O.-G. and M.L.F.-M. Drafting the work and revising it critically for important intellectual content: J.M.M.-A. and M.L.F.-M. Final approval of the submitted version: J.M.M.-A. and M.L.F.-M. Agreement to be accountable for all aspects of the work in ensuring that questions related to the accuracy or integrity of any part of the work are appropriately investigated and resolved: J.M.M.-A. and M.L.F.-M. All authors have read and agreed to the published version of the manuscript.

Funding: This research received no external funding.

Acknowledgments: The first author thanks the Banco de Santander for granting a USC-Banco Santander fellowship, which has made possible the completion of this work. The authors thank Timoteo Caetano Ferreira for revising the manuscript and three anonymous reviewers for their valuable suggestions and constructive criticism.

Conflicts of Interest: The authors declare no conflict of interest.

References

1. IUSS Working Group WRB. *World Reference Base for Soil Resources 2014, Update 2015, International Soil Classification System for Naming Soils and Creating Legends for Soil Maps*; FAO: Rome, Italy, 2015.
2. Effland, W.; Pouyat, R. The genesis, classification, and mapping of soils in urban areas. *Urban Ecosyst.* **1997**, *1*, 217–228. [CrossRef]
3. Morel, J.L.; Chenu, C.; Lorenz, K. Ecosystem services provided by soils of urban, industrial, traffic, mining, and military areas (SUITMAs). *J. Soils Sediments* **2015**, *15*, 1659–1666. [CrossRef]
4. Sharma, R.K.; Agrawal, M.; Marshall, F. Heavy metal contamination of soil and vegetables in suburban areas of Varanasi, India. *Environ. Saf.* **2007**, *66*, 258–266. [CrossRef]
5. Iram, S.; Ahmad, I.; Akhtar, S. Distribution of Heavy Metals in Peri-Urban Agricultural Areas Soils. *J. Chem. Soc. Pak.* **2012**, *34*, 861–869.
6. Sun, Y.B.; Zhou, Q.X.; Xie, X.K.; Liu, R. Spatial, sources and risk assessment of heavy metal contamination of urban soils in typical regions of Shenyang, China. *J. Hazard. Mater* **2010**, *174*, 455–462. [CrossRef]
7. Huang, Y.; Li, T.; Wu, C.; He, Z.; Japenga, J.; Deng, M.; Yang, X. An integrated approach to assess heavy metal source apportionment in peri-urban agricultural soils. *J. Hazard. Mater* **2015**, *299*, 540–549. [CrossRef]
8. Papa, S.; Bartoli, G.; Pellegrino, A.; Fioretto, A. Microbial activities and trace element contents in an urban soil. *Environ. Monit. Assess.* **2010**, *165*, 193–203. [CrossRef]
9. Kumar, K.; Hundal, L.S. Soil in the City: Sustainably Improving Urban Soils. *J. Environ. Qual.* **2016**, *45*, 2–8. [CrossRef]
10. Pouyat, R.V.; Yesilonis, I.D.; Russell-Anelli, J.; Neerchal, N.K. Soil Chemical and Physical Properties That Differentiate Urban Land-Use and Cover Types. *Soil Sci. Soc. Am. J.* **2007**, *71*, 1010–1019. [CrossRef]
11. Massas, I.; Kalivas, D.; Ehaliotis, C.; Gasparatos, D. Total and available heavy metal concentrations in soils of the Thriassio plain (Greece) and assessment of soil pollution indexes. *Environ. Monit. Assess.* **2013**, *185*, 6751–6766. [CrossRef]
12. Gasparatos, D.; Mavromati, G.; Kotsovilis, P.; Massas, I. Fractionation of heavy metals and evaluation of the environmental risk for the alkaline soils of the Thriassio plain: A residential, agricultural, and industrial area in Greece. *Environ. Earth Sci.* **2015**, *74*, 1099–1108. [CrossRef]
13. Rodríguez-Seijo, A.; Andrade, M.L.; Vega, F.A. Origin and spatial distribution of metals in urban soils. *J. Soils Sediments* **2017**, *17*, 1514–1526. [CrossRef]
14. Manta, D.S.; Angelone, M.; Bellanca, A.; Neri, R.; Sprovieri, M. Heavy metals in urban soils: A case study from the city of Palermo (Sicily), Italy. *Sci. Total. Environ.* **2002**, *300*, 229–243. [CrossRef]

15. Massas, I.; Ehaliotis, C.; Kalivas, D.; Panagopoulou, G. Concentrations and Availability Indicators of Soil Heavy Metals; the Case of Children's Playgrounds in the City of Athens (Greece). *Water Air Soil Pollut.* **2010**, *212*, 51–63. [CrossRef]
16. Lee, C.S.-l.; Li, X.; Shi, W.; Cheung, S.C.-n.; Thornton, I. Metal contamination in urban, suburban, and country park soils of Hong Kong: A study based on GIS and multivariate statistics. *Sci. Total. Environ.* **2006**, *356*, 45–61. [CrossRef]
17. Delbecque, N.; Verdoodt, A. Spatial Patterns of Heavy Metal Contamination by Urbanization. *J. Environ. Qual.* **2016**, *45*, 9–17. [CrossRef]
18. National Bureau of Statistics and Information. Statistical Yearbook of Cuba 2018. 2019. Available online: http://www.one.cu/aec2018.htm (accessed on 23 July 2019).
19. Companioni, N.; Rodríguez-Nodals, A.; Sardiñas, J. Avances de la agricultura urbana, suburbana y familiar. *Agroecología* **2017**, *12*, 91–98.
20. Koont, S. A Cuban Success Story: Urban Agriculture. *Rev. Radic. Politi- Econ.* **2008**, *40*, 285–291. [CrossRef]
21. Orsini, F.; Kahane, R.; Nono-Womdim, R.; Gianquinto, G. Urban agriculture in the developing world: A review. *Agron. Sustain. Dev.* **2013**, *33*, 695–720. [CrossRef]
22. Wright, J. *Sustainable Agriculture and Food Security in an Era of Oil Scarcity: Lessons from Cuba*; Routledge: London, UK, 2012; p. 280.
23. USEPA. Method 3051A. Microwave Assisted Acid Digestion of Sediments, Sludges, Soils, and Oils. Available online: https://www.epa.gov/hw-sw846/sw-846-test-method-3051a-microwave-assisted-acid-digestion-sediments-sludges-soils-and-oils (accessed on 11 May 2020).
24. Mehlich, A. Mehlich 3 soil test extractant: A modification of Mehlich 2 extractant. *Commun. Soil Sci. Plant Anal.* **1984**, *15*, 1409–1416. [CrossRef]
25. Schwertmann, U. The differentiation of iron oxide in soils by a photochemical extraction with acid ammonium oxalate. *Z. Pflanz. Düngung Bodenkd.* **1964**, *105*, 194–202. [CrossRef]
26. Rodriguez Alfaro, M.; Montero, A.; Muniz Ugarte, O.; Araujo do Nascimento, C.W.; de Aguiar Accioly, A.M.; Biondi, C.M.; Agra Bezerra da Silva, Y.J. Background concentrations and reference values for heavy metals in soils of Cuba. *Environ. Monit. Assess.* **2015**, *187*. [CrossRef]
27. Kabata-Pendias, A. *Trace Elements in Soils and Plants*; CRC: Boca Raton, FL, USA, 2011.
28. Ajmone-Marsan, F.; Biasioli, M. Trace Elements in Soils of Urban Areas. *Water Air Soil Pollut.* **2010**, *213*, 121–143. [CrossRef]
29. Madrid, F.; Biasioli, M.; Ajmone-Marsan, F. Availability and bioaccessibility of metals in fine particles of some urban soils. *Arch. Environ. Contam. Toxicol.* **2008**, *55*, 21–32. [CrossRef]
30. Wei, B.G.; Yang, L.S. A review of heavy metal contaminations in urban soils, urban road dusts and agricultural soils from China. *Microchem. J.* **2010**, *94*, 99–107. [CrossRef]
31. Diaz Rizo, O.; Echeverria Castillo, F.; Arado Lopez, J.O.; Hernandez Merlo, M. Assessment of Heavy Metal Pollution in Urban Soils of Havana City, Cuba. *Bull. Environ. Contam. Toxicol.* **2011**, *87*, 414–419. [CrossRef]
32. Senesi, G.S.; Baldassarre, G.; Senesi, N.; Radina, B. Trace element inputs into soils by anthropogenic activities and implications for human health. *Chemosphere* **1999**, *39*, 343–377. [CrossRef]
33. Liu, L.; Liu, Q.; Ma, J.; Wu, H.; Qu, Y.; Gong, Y.; Yang, S.; An, Y.; Zhou, Y. Heavy metal(loid)s in the topsoil of urban parks in Beijing, China: Concentrations, potential sources, and risk assessment. *Environ. Pollut.* **2020**, *260*. [CrossRef]
34. Ajmone-Marsan, F.; Padoan, E.; Madrid, F.; Vrscaj, B.; Biasioli, M.; Davidson, C.M. Metal Release under Anaerobic Conditions of Urban Soils of Four European Cities. *Water Air Soil Pollut.* **2019**, *230*. [CrossRef]
35. Franco-Uria, A.; Lopez-Mateo, C.; Roca, E.; Fernandez-Marcos, M.L. Source identification of heavy metals in pastureland by multivariate analysis in NW Spain. *J. Hazard. Mater.* **2009**, *165*, 1008–1015. [CrossRef]
36. Nriagu, J.O. Natural versus anthropogenic emissions of trace-metals to the atmosphere. In *Control and Fate of Atmospheric Trace Metals*; Pacyna, J.M., Ottar, B., Eds.; Springer: Berlin/Heidelberg, Germany, 1989; Volume 268, pp. 3–13.
37. Rodríguez-Mercado, J.J.; Altamirano-Lozano, M.A. Vanadio: Contaminación, Metabolismo y Genotoxicidad. *Rev. Int. Contam. Ambient.* **2006**, *22*, 173–189.
38. Diaz Rizo, O.; Fonticiella Morell, D.; Arado Lopez, J.O.; Borrell Munoz, J.L.; D'Alessandro Rodriguez, K.; Lopez Pino, N. Spatial Distribution and Contamination Assessment of Heavy Metals in Urban Topsoils from Las Tunas City, Cuba. *Bull. Environ. Contam. Toxicol.* **2013**, *91*, 29–35. [CrossRef]

39. Monterroso, C.; Alvarez, E.; Marcos, M.L.F. Evaluation of Mehlich 3 reagent as a multielement extractant in mine soils. *Land Degrad. Dev.* **1999**, *10*, 35–47. [CrossRef]
40. Wendt, J.W. Evaluation of the Mehlich 3 soil extractant for upland Malawi soils. *Commun. Soil Sci. Plant Anal.* **1995**, *26*, 687–702. [CrossRef]
41. Baken, S.; Larsson, M.A.; Gustafsson, J.P.; Cubadda, F.; Smolders, E. Ageing of vanadium in soils and consequences for bioavailability. *Eur. J. Soil Sci.* **2012**, *63*, 839–847. [CrossRef]
42. Jeske, A.; Gworek, B. Chromium, nickel and vanadium mobility in soils derived from fluvioglacial sands. *J. Hazard. Mater.* **2012**, *237*, 315–322. [CrossRef]
43. Larsson, M.A.; Baken, S.; Gustafsson, J.P.; Hadialhejazi, G.; Smolders, E. Vanadium bioavailability and toxicity to soil microorganisms and plants. *Environ. Toxicol. Chem.* **2013**, *32*, 2266–2273. [CrossRef]
44. Caridad-Cancela, R.; Paz-González, A.; de Abreu, C.A. Total and Extractable Nickel and Cadmium Contents in Natural Soils. *Commun. Soil Sci. Plant Anal.* **2005**, *36*, 241–252. [CrossRef]
45. Rodriguez Alfaro, M.; Araujo do Nascimento, C.W.; Muniz Ugarte, O.; Montero Alvarez, A.; de Aguiar Accioly, A.M.; Calero Martin, B.; Limeres Jimenez, T.; Ginebra Aguilar, M. First national-wide survey of trace elements in Cuban urban agriculture. *Agron. Sustain. Dev.* **2017**, *37*. [CrossRef]
46. Agbenin, J.O.; Welp, G.; Danko, M. Fractionation and Prediction of Copper, Lead, and Zinc Uptake by Two Leaf Vegetables from Their Geochemical Fractions in Urban Garden Fields in Northern Nigeria. *Commun. Soil Sci. Plant Anal.* **2010**, *41*, 1028–1041. [CrossRef]
47. Shaheen, S.M.; Rinklebe, J.; Rupp, H.; Meissner, R. Temporal dynamics of pore water concentrations of Cd, Co, Cu, Ni, and Zn and their controlling factors in a contaminated floodplain soil assessed by undisturbed groundwater lysimeters. *Environ. Pollut.* **2014**, *191*, 223–231. [CrossRef]
48. Florido, M.C.; Madrid, F.; Ajmone-Marsan, F. Variations of Metal Availability and Bio-accessibility in Water-Logged Soils with Various Metal Contents: In Vitro Experiments. *Water Air Soil Pollut.* **2011**, *217*, 149–156. [CrossRef]

© 2020 by the authors. Licensee MDPI, Basel, Switzerland. This article is an open access article distributed under the terms and conditions of the Creative Commons Attribution (CC BY) license (http://creativecommons.org/licenses/by/4.0/).

Article

Measuring Soil Metal Bioavailability in Roadside Soils of Different Ages

Shamali De Silva [1,*], Trang Huynh [2], Andrew S. Ball [3], Demidu V. Indrapala [1,3] and Suzie M. Reichman [1,4]

1. School of Engineering, RMIT University, Melbourne 3001, Australia; s3786196@student.rmit.edu.au (D.V.I.); suzie.reichman@unimelb.edu.au (S.M.R.)
2. Hydrobiology Pty Ltd., Auchenflower 4066, Australia; trang.huynh@hydrobiology.com
3. School of Science, RMIT University, Melbourne 3001, Australia; andy.ball@rmit.edu.au
4. Centre for Anthropogenic Pollution Impact and Management, School of Biosciences, University of Melbourne, Parkville 3010, Australia
* Correspondence: shamali.desilva@rmit.edu.au

Received: 16 August 2020; Accepted: 11 October 2020; Published: 15 October 2020

Abstract: Finding a reliable method to predict soil metal bioavailability in aged soil continues to be one of the most important problems in contaminated soil chemistry. To investigate the bioavailability of metals aged in soils, we used roadside soils that had accumulated metals from vehicle emissions over a range of years. We collected topsoil (0–10 cm) samples representing new-, medium- and old-aged roadside soils and control site soil. These soils were studied to compare the ability of the diffusive gradients in thin films technique (DGT), soil water extraction, $CaCl_2$ extraction, total metal concentrations and optimised linear models to predict metal bioavailability in wheat plants. The response time for the release of metals and the effect on metal bioavailability in field aged soils was also studied. The DGT, and extractable metals such as $CaCl_2$ extractable and soil solution metals in soil, were not well correlated with metal concentrations in wheat shoots. In comparison, the strongest relationships with concentrations in wheat shoots were found for Ni and Zn total metal concentrations in soil (e.g., Ni r = 0.750, p = 0.005 and Zn r = 0.833, p = 0.001); the correlations were still low, suggesting that total metal concentrations were also not a robust measure of bioavailability. Optimised linear models incorporating soil physiochemical properties and metal extracts together with road age as measure of exposure time, demonstrated a very strong relationship for Mn R^2 = 0.936; Ni R^2 = 0.936 and Zn R^2 = 0.931. While all the models developed were dependent on total soil metal concentrations, models developed for Mn and Zn clearly demonstrated the effect of road age on metal bioavailability. Therefore, the optimised linear models developed have the potential for robustly predicting bioavailable metal concentrations in field soils where the metals have aged in situ. The intrinsic rate of release of metals increased for Mn (R^2 = 0.617, p = 0.002) and decreased for Cd (R^2 = 0.456, p = 0.096), Cu (R^2 = 0.560, p = 0.083) and Zn (R^2 = 0.578, p = 0.072). Nickel did not show any relationship between dissociation time (Tc) and road age. Roadside soil pH was likely to be the key parameter controlling metal aging in roadside soil.

Keywords: vehicular emissions; road age; diffusive gradients in thin films (DGT); metal dissociation time (Tc); wheat assay; optimised linear model

1. Introduction

Metals are persistent, bioaccumulate and are well-known for their toxicity. The presence of elevated concentrations of metals in soil due to natural or anthropogenic activity represents a potential risk to human and ecological health; consequently, the assessment of soil metal contamination risks is of great interest to governments and regulators [1]. The measurement and understanding of the bioavailability of elevated metals in soil are an important part of any risk assessment. The bioavailability of metals in soil is a complex dynamic process and believed to be driven by speciation, sorption and biological processes in soil, but still little is known about their interplay in soil systems [2]. Previous studies have shown the importance of soil metal bioavailability in determining metal toxicity; however, there is still no general agreement on how best to measure metal bioavailability in soils [2–6]. Over the years, researchers have attempted to measure bioavailable metals in soil using many techniques such as ion exchange extraction, chelate extractions, rhizosphere extractions, isotopic solution extractions and the diffusive gradients in thin films technique (DGT). Effective elemental concentrations available for uptake from the solution-phase and solid-phase [7] can be simulated using the 2D DIFS model used in DGT. The DGT technique has also been used to estimate metal release kinetics from the solid to solution phase in soils [8,9]. The response time of soil to the resupply of metals in to the solution phase represents the response to metal depletion, and is directly related to the rate constant of the metal resupply process from the solution phase to the DGT film and therefore to plant roots [7]. To predict the concentration of metals in wheat plants as a result of the uptake of metals from uncontaminated soil, mathematical models including multiple regression analysis have also been validated and applied [10,11]. Generally, these techniques may not account for metals that have been accumulated over time in soil i.e., the aging effect of metals in soil. Thus, there is still a need to find and validate techniques for measuring bioavailable metals in aged contaminated soil.

Metal aging in soil is the process by which the bioavailability, mobility and/or the exchangeability of metals declines over time. The aging process is also referred to as natural attenuation or the fixation effect of metals in soil [12]. Aging of metals in soil is a long-term process and an important aspect affecting metal bioavailability [13,14]. However, our current understanding of the kinetics of metal aging in soil is limited. Experiments have confirmed that metal bioavailability in soil decreases as a result of the aging process [15–17]. Temperature, pH, moisture content and climate affect the rate of metal aging in soil, with pH considered as the most important parameter [18]. However, the majority of experiments on metal aging have been conducted using soils that have been spiked with metals in a laboratory [19], and thus have only assessed short-term aging effects.

The aim of this study was to systematically investigate metal bioavailability in roadside soil at varying road ages and metal concentrations. Wheat plants were used as a bioassay to accumulate bioavailable metals. The ability of DGT, soil water extraction, $CaCl_2$ extraction, total metal concentrations in soil and optimised linear models to measure metal bioavailability, together with the release response time of metals and its impact on metal bioavailability in field-aged soils were studied. We used roadside soil to investigate soil properties and metal aging effects on bioavailability for a wide range of metals. Cadmium, Cu, Ni, Mn and Zn have previously been found to be elevated in roadside soils from Melbourne, Australia [20]. More information on the vehicular sources of Cd, Cu, Mn, Ni and Zn in roadside soils can be found in De Silva et al. [21]. Soil samples were collected from sites representing three different road ages, i.e., new, medium and old-aged roads.

2. Methods

2.1. Soil Sampling

Metals in roadside soils have been deposited from vehicle emissions over varying time periods of up to multiple decades depending on the age of the roads. Thus, roadside soils provide an opportunity to study bioavailability and metal aging processes in long-term in situ metal-contaminated soils. Soil samples were collected in December 2012 from roadside soils formed on chromosol [22] of newer volcanic basaltic geology [23] in the west of Melbourne, Australia. The selection of roads from the same geology ensured that the soil derived from similar parent material reduced variations due to natural soil derivatives. All chosen roads were constructed from the same surface material, i.e., asphalt. Samples were collected adjacent to roads carrying ≥ 1500 vehicles/day and represented three different road ages, i.e., new (N = 1.5–5 years), medium (M = 5–10 years) and old (O ≥ 15 years). Three soil replicates for each road type were collected. Three sites in parklands at least 1 km away from roads and industry were included as control sites. Approximately 500 g of topsoil samples from 0–10 cm were collected using a clean stainless steel spatula and placed in zip-lock polythene bags. The immediate road edge up to 1 m was avoided to minimise the risk of sampling refilled or recently disturbed soil from road constructions and upgrades [24]. Soil samples were air-dried (23 ± 2 °C), sieved (≤2 mm), sealed in zip lock polythene bags and stored at ambient laboratory conditions (20 ± 2 °C) in the dark until further analysis [20].

2.2. Wheat Assay

Soil aliquots (75 g) were placed into 24 polystyrene plastic pots (180 mL). The pots were moistened to 50% of the maximum water holding capacity (MWHC) with ultrapure water (18 MΩ.cm) and left to equilibrate for 24 h at 21 °C in the dark. *Triticum aestivum* cv. AXE (bread wheat) seeds (Department of Primary Industries, New South Wales, Australia) were surface sterilised with 0.3% NaOCl solution and rinsed with deionised water [25]. Twenty seeds were sown into each pot and the pots placed in a growth chamber at 21 °C/15 °C on a 14 h/10 h light/dark cycle for 28 days. On the 5th day after sowing, the seedlings were thinned to 10 plants per pot. Aliquots (5 mL) of Ruakara nutrient solution [6] were added to the pots on days 5, 10, 15, 20 and 25 prior to watering. Throughout the experiment, pots were regularly watered with ultrapure water to maintain 50% MWHC.

On day 28, seedlings were harvested by cutting shoots 1 cm above the soil surface. Shoots were washed thoroughly with ultrapure water, dried for 24 h at 45 °C, and the shoot dry weight was recorded for each pot. Dried shoot samples were finely ground and stored in polypropylene plastic storage vials at ambient laboratory conditions (20 ± 2 °C) in the dark until further processing. Nitric acid (HNO_3, 5 mL, 70%, Sigma Aldrich analytical grade), together with hydrogen peroxide (H_2O_2, 1 mL, 30% *w/w*, Sigma Aldrich analytical grade), was added to aliquots of the dried plant samples (0.2 g) and digested at 115 °C using a microwave digester (Milestone, Ethos d' model). After digestion, samples were filtered (pore size 0.45 μm) and diluted with ultrapure water before being analysed for total metals using inductively coupled plasma-mass spectrometry (ICP-MS, Agilent Technologies 7700x element analyser). The isotopes chosen for this investigation were Cd (111), Cu (63), Ni (60), Mn (55) and Zn (66). The accuracy of the method was verified by analysing reagent blanks, which represented 10% of the total digested sample population and replicates of certified reference citrus plant material [NCS ZC73018 (GSB-11)]. The recovery of the certified reference material was >90% for all certified metals and no signs of contamination were found.

2.3. Analysis of Metals in Soil

2.3.1. Total Metal Digestion

Air-dried and sieved soil aliquots of 0.5 g were digested in 2.5 mL of *aqua regia* (HNO_3: HCl, 1:3) at 105 °C for 2 h, followed by the addition of 1 mL of H_2O_2 30% (*w/w*). Digested soil samples were cooled and final volumes were diluted to 50 mL using ultrapure water. Diluted acid digestate was then filtered using 0.45 μm nylon filter membranes. To determine the contamination, precision and bias of the analysis, reagent blanks which represented 10% of the total digested sample population and replicated certified reference materials (DO82-540, ERA) were incorporated into the analysis. The findings of the analytical results showed no signs of contamination, and the accuracy was found to be consistently within 10% of the certified values [20].

2.3.2. Metal $CaCl_2$ Extraction

For the extraction of soluble metals from soil, air-dried soil aliquots (10 g) were placed in 250 mL high density polyethylene (HDPE) plastic bottles. Calcium chloride (100 mL, 0.01 M) was then added to each bottle [26]. The bottles were vortexed for 2 h at 20 °C. The suspension was decanted into a 100 mL centrifuge tube and centrifuged for 10 min at 1800 rpm. A 10 mL aliquot of the supernatant was filtered using a 0.45 µm syringe filter. The filtered supernatant was acidified with 4% HNO_3 at a ratio of 1:2 ratio before metal analysis.

2.3.3. Analysis of Metals in Soil Digests and Extracts

Total metals and $CaCl_2$ extractable metals were analysed for Cd, Cu, Mn, Ni and Zn using ICP-MS. The isotopes chosen for this investigation were Cd (111), Cu (63), Ni (60), Mn (55) and Zn (66). The accuracy of the method was verified by analysing certified reference soil (DO82-540) alongside samples and with recovery >90% for all certified metals as previously described in De Silva et al. [20].

2.3.4. Other Soil Analysis

Methods described in Rayment and Higginson [27] were used to determine soil parameters. The following soil analyses methods were used: pH (method 4A1) using 1:5 soil/water extract, effective cation exchange capacity (ECEC) (method 15D1) using 1 M ammonium acetate at pH 7.0; pre-treatment for soluble salts and manual leaching, and organic carbon (OC) (method 6A1) using the Walkley and Black (W and B) method.

2.4. Diffusive Gradients in Thin Films (DGT) Technique

The DGT soil measurement method followed the DGT soil deployment protocol guided by DGT research (DGT soil deployment guide A2-20), and as described by Zhang et al. [28]. Briefly, 50 g soil aliquots were moistened to MWHC by adding the measured mass of ultrapure water while mixing with a wooden spatula until a smooth paste was formed. The soil paste was covered with clean plastic wrap to prevent water evaporation and left to equilibrate for 24 h in the dark at 20 ± 1 °C. Before DGT deployment in the soil, the DGT (Chelex-100) devices (Griffith University, Australia) were rinsed thoroughly with ultrapure water to remove the $NaNO_3$ storage solution and dried with tissue paper. A small subsample of each soil paste was applied directly to the DGT device membrane to ensure complete contact between the soil paste and the DGT membrane, before the device was placed with gentle pressure directly onto the soil paste sample. During deployment of the DGT devices, the dishes were covered with plastic paraffin film (Parafilm M®, Bemis NA) to minimise water evaporation. Blank DGT without soil paste were also deployed. In this experiment, the deployment time of 7 h was used to avoid exceeding the capacity of the resin gel, as verified by calculations for the highest concentrations of metals in water extractable metal concentrations [29,30]. The deployment occurred at ambient laboratory lighting conditions. The temperature throughout the deployment was recorded using two data loggers and a mean temperature of deployment period of 20 °C (range 19–21 °C) was used for DGT calculation.

Upon retrieval, each DGT device was rinsed using ultrapure water to remove soil particles, and dried using clean, dry tissue paper. The resin gel was removed using separate, clean plastic tweezers for each sample. Metals bound within the resin gel were extracted using 1 mL of 1 M HNO_3 in 5 mL polypropylene centrifuge tubes. To obtain soil solution samples for the analysis of soil solution metal concentration (CSOL), an aliquot of the sample of the soil paste was added to a 50 mL centrifuge tube and centrifuged at 5000 rpm for 5 min. The supernatant was decanted and filtered through a nylon filter membrane (0.45 µm). The metal concentrations in the resin gel and soil solution samples were analysed using ICP-MS.

DGT Data Analysis

The concentration of metal accumulated on the binding gel in the DGT device (M) was determined according to Equation (1).

$$M = C\ (V_{acid} + V_{gel})/fe \qquad (1)$$

where: C is the metal concentration eluted from the binding gel measured by ICP-MS

V_{acid} is the volume of acid used for elution (V_{acid} = 1 mL),

V_{gel} is the volume of resin gel (V_{gel} = 0.16 mL),
fe is the elution factor (fe = 0.8).

Once M was calculated, the interfacial DGT concentration (C_{DGT}) was calculated using Equation (2),

$$C_{DGT} = M\Delta g/DAt \qquad (2)$$

where: Δg is the diffusive layer thickness (0.8 mm) plus the thickness of the filter membrane (0.14 mm), which is 0.94 mm,

D is the diffusion coefficient of metal at a given temperature (cm^2 S^{-1}),

A is the area of the exposed membrane (A = 3.14 cm^2),

t is the deployment time (in seconds).

The diffusion coefficients of the metal of interest (Ds) were calculated using Equations (3)–(5).

$$Pc = m/V \qquad (3)$$

$$\phi = Dp / (Pc + Dp) \qquad (4)$$

$$Ds = D_0 / (1 - \ln \phi^2) \qquad (5)$$

where: m is the total mass of all soil particles;

V the pore water volume in a given volume of total soil (cm^3),

Dp is the density of soil particles (2.65 g cm^{-3}) in soil,

Do is the diffusion coefficient of the metal ion at 20 ± 1 °C, (cm^2 S^{-1}),

Ds is the diffusion coefficient in sediment (cm^2 S^{-1}).

The input parameters used in the 2D DFIS model to calculate R_{DIFF}, were particle concentration (Pc) and soil porosity (ϕ). Pc (g cm^{-3}) was determined using Equation (3) and soil porosity (ϕ) was determined using Equation (4), and the diffusion coefficient in sediment (Ds) was calculated using Equation (5).

Effective solution concentrations, C_E, were derived using Equation (6).

$$C_E = C_{DGT} / R_{DIFF} \qquad (6)$$

where: C_{DGT} was the metal concentration measured by the DGT technique (in mg/kg).

The effective concentration (C_E) was used to estimate the potential metal concentration that can be absorbed by plants. The DGT technique is designed to mimic root uptake metal from soil. During DGT measurement, interface metal concentrations were measured to take into account the continuous depletion of metals, due to the uptake and the resupply process of metals from solid to solution phase. Metal depletion is indicated as a ratio (R) using the C_{DGT} and independently measured solution (C_{SOL}) concentrations (Equation (7)).

$$R = C_{DGT} / C_{SOL} \qquad (7)$$

$$Tc = C (1 - R/R - d)^2 \qquad (8)$$

where: Tc (seconds) is the time taken to reach the equilibrium, Tc is determined using Equation (8).

Note: For samples with high porosity; C = 403, d = 0.0247 and for low porosity case C = 229, d = 0.0186 [7]. The soils for this study were treated as being low porosity.

2.5. Statistical Analysis

Data were analysed using SPSS Statistics IBM version 21 (2012). Soil physicochemical parameters were analysed for descriptive statistics (mean ± standard error). Roadside and control sites were compared using one-way ANOVA, with significance taken as $p < 0.05$. Where necessary, data were square root or $\log_{10}$ transformed to obtain an acceptable normal distribution and to stabilise the variance of the residuals, and significance was based on an analysis of the transformed data. Relationships between the metal of interest and other physicochemical parameters were analysed using a 2-tailed Pearson correlation test; significance was treated as $p < 0.05$. Optimised linear models for soil metal concentrations were determined using stepwise regression with criteria of $\alpha = 0.05$ for entry. The input parameters used were soil physiochemical properties including pH, EC, ECEC, TOC, Tc, DGT, soil total metals, $CaCl_2$ extractable metals, water extractable metals, and the age of the road as a measure of metal aging in soil. Linear correlations were also developed; $p < 0.05$ was considered significant and $0.05 < p < 0.10$ was considered marginally significant.

3. Results

3.1. Soil Properties

There were differences in general soil characteristics between road types and the control sites (Table 1). Significantly more alkaline pH was recorded from new roadsides compared to other sites; total organic carbon was highest in the oldest roadside soils and EC was lowest in the control soil. There was no significant difference in Pc or ECEC between the roadside soil sites, i.e., new, medium and old roads.

Table 1. Soil characteristics for roadside soils in western Melbourne, Australia, on Newer Volcanic parent material; values are mean ± standard error (n = 3). Soils were collected from new (N = 1.5–5 years), medium (M = 5–10 years) and old (O ≥ 15 years) roads. Numbers in a column with the same letter are not significantly different ($p > 0.05$).

Site	pH	Pc (g/cm^3)	TOC (C% dry Soil)	EC (mS/cm)	ECEC (meq/100 g)
Control	5.2 ± <0.1 a	1.6 ± <0.1 a	1.8 ± 0.1 a	48 ± 4 a	8 ± 2 a
New	8.2 ± 0.1 b	1.6 ± 0.3 a	1.8 ± 0.4 a	174 ± 37 b	31 ± 3 a
Medium	6.2 ± 0.2 ab	1.4 ± 0.4 a	1.9 ± 0.5 a	225 ± 39 b	25 ± 3 a
Old	6.3 ± 0.5 ab	1.6 ± <0.1 a	3.5 ± 0.7 b	162 ± 15 b	22 ± 2 a

Pc = soil particle concentration determined using Equation (3); EC = electrical conductivity, TOC = total organic carbon, ECEC = effective cation exchange capacity. Values are means (n = 3), ± standard error, a, b, c, denotes the significance between the sites at $p < 0.05$.

3.2. Metal Concentrations in Soil, Wheat, DGT, C_E, C_{SOL}, $CaCl_2$ Extractable Metals

Total metal (Cd, Cu, Mn, Ni and Zn) concentrations measured in the control soil were lower than in the roadside soil (Table 2). The highest Zn and Cd concentrations (45 mg/kg and 0.2 mg/kg) were recorded adjacent to old roads and the highest Mn concentration (599 mg/kg) was recorded next to a new road. The highest Cu concentration (12 mg/kg) and highest Ni concentration (21 mg/kg) were recorded from medium-aged roads [20].

Table 2. Metal concentrations (Cd, Cu, Mn, Ni and Zn) in soil, wheat, DGT, C_E, C_{SOL}, $CaCl_2$ extractable metals for soil and soil after the growth of wheat.

Metal	Type	Total (mg/kg)	$CaCl_2$ (µg/L)	C_{SOL} (µg/L)	C_E (µg/L)	C_{DGT} (µg/L)	Plant (mg/kg DW)	R
Cd	C	0.05 ± 0.02 a	0.22 ± 0.01 a	0.004 ± <0.001 a	0.02 ± 0.02 a	0.002 ± 0.001 a	0.02 ± 0.01 a	0.41 ± 0.17 b
	N	0.07 ± 0.02 ab	0.10 ± <0.01 a	0.011 ± <0.001 a	0.02 ± 0.01 a	0.002 ± 0.001 a	0.04 ± 0.03 ab	0.09 ± 0.05 a
	M	0.13 ± 0.01 b	0.30 ± 0.032 a	0.008 ± <0.001 a	0.06 ± 0.02 a	0.005 ± 0.000 a	0.07 ± 0.01 b	0.39 ± 0.09 b
	O	0.21 ± 0.02 b	0.16 ± <0.01 a	0.027 ± 0.010 a	0.07 ± 0.02 a	0.006 ± 0.002 a	0.01 ± 0.00 a	0.11 ± 0.05 a
Cu	C	3.94 ± 0.45 a	0.02 ± 0.01 a	0.03 ± 0.006 a	0.04 ± 0.01 a	0.003 ± 0.001 a	1.78 ± 0.00 a	0.13 ± 0.01 a
	N	7.80 ± 0.98 b	0.08 ± 0.03 ab	0.04 ± 0.006 a	0.03 ± 0.01 a	0.002 ± 0.001 a	1.87 ± 0.36 a	0.05 ± 0.01 a
	M	8.96 ± 1.16 b	0.17 ± 0.12 b	0.06 ± 0.02 a	0.10 ± 0.03 b	0.009 ± 0.003 ab	2.36 ± 1.18 a	0.14 ± 0.02 a
	O	8.65 ± 0.65 b	0.06 ± 0.01 ab	0.07 ± 0.02 a	0.11 ± 0.05 b	0.010 ± 0.004 b	2.38 ± 1.00 a	0.13 ± 0.02 a
Mn	C	55 ± 12 a	0.02 ± 0.01 a	0.53 ± 0.11 a	2.10 ± 0.07 a	0.21 ± 0.01 a	14.37 ± 9.79 a	0.42 ± 0.08 a
	N	599 ± 26 c	0.01 ± 0.004 a	0.21 ± 0.07 a	1.29 ± 0.58 a	0.12 ± 0.06 a	52.88 ± 2.88 b	0.53 ± 0.10 a
	M	323 ± 117 b	0.03 ± 0.03 a	0.37 ± 0.02 a	2.28 ± 0.37 a	0.20 ± 0.04 a	71.50 ± 16.2 b	0.53 ± 0.08 a
	O	171 ± 54 b	0.06 ± 0.002 a	0.80 ± 0.26 b	3.32 ± 0.11 a	0.27 ± 0.01 a	59.84 ± 7.94 b	0.54 ± 0.07 a

Table 2. Cont.

Metal	Type	Total (mg/kg)	µg/L				Plant (mg/kg DW)	R
			CaCl$_2$	C$_{SOL}$	C$_E$	C$_{DGT}$		
Ni	C	4.3 ± 1.20 a	0.14 ± 0.03 a	0.016 ± 0.006 a	0.069 ± 0.02 a	0.006 ± 0.001 a	2.70 ± 0.61 a	0.25 ± 0.06 a
	N	16.6 ± 5.7 b	0.07 ± 0.01 a	0.054 ± 0.011 a	0.065 ± 0.005 a	0.006 ± 0.001 a	10.5 ± 4.98 b	0.12 ± 0.01 a
	M	19.0 ± 3.9 b	0.15 ± 0.05 a	0.053 ± 0.005 a	0.118 ± 0.019 a	0.011 ± 0.001 b	9.09 ± 0.43 b	0.21 ± 0.01 a
	O	14.5 ± 1.1 b	0.13 ± 0.03 a	0.058 ± 0.012 a	0.106 ± 0.016 a	0.010 ± 0.001 b	10.81 ± 2.31 b	0.18 ± 0.01 a
Zn	C	8.4 ± 1.8 a	2.20 ± 0.69 a	0.04 ± 0.301 a	0.36 ± 0.06 a	0.034 ± 0.006 a	06.30± 1.74 a	0.20 ± 0.08 a
	N	25.7 ± 6.7 b	1.48 ± 0.48 a	1.14 ± 0.115 b	0.21 ± 0.03 a	0.02 ± 0.003 a	24.66 ± 1.91 b	0.02 ± 0.00 a
	M	38.6 ± 3.9 b	2.97 ± 0.32 a	0.43 ± 0.301 ab	0.38 ± 0.06 a	0.036 ± 0.005 a	25.34 ± 2.51 b	0.19 ± 0.07 a
	O	45.2 ± 1.4 b	2.10 ± 0.87 a	1.02 ± 0.004 b	0.54 ± 0.16 a	0.052 ± 0.015 a	28.47 ± 1.73 b	0.05 ± 0.02 a

DGT = Diffusive gradients in thin film, CE = effective concentration, C$_{SOL}$ = pore water metal concentrations, CaCl$_2$ extractable = available metals, R = ratio of C$_{DGT}$ and C$_{SOL}$, C = soil not exposed to vehicular emissions (control), N = new aged roads (2–5 years), M = medium aged roads (5–10 years) and O = old aged roads (≥15 years). Values are means ($n = 3$), ± standard error, a, b, c, denotes the significance between the sites at $p < 0.05$.

Metal concentrations in wheat shoots varied across the sites. Significantly higher wheat shoot concentrations were measured for Mn ($p = 0.041$), Ni ($p = 0.024$) and Zn ($p = 0.035$) for roadside soils compared to the control sites (Table 2). Cadmium showed a significant accumulation in wheat shoots grown in medium-aged roadside soil compared to control site grown wheat. Copper accumulation did not vary significantly in wheat shoots at any site. Some roadside soil samples showed a significant difference in CaCl$_2$ extracted metals, soil solution metals (C$_{SOL}$) and effective concentrations (C$_E$) among roadside and control soils (Table 2).

Both the DGT technique and CaCl$_2$ extractable metals were not statistically correlated to metals in wheat shoots for plants grown in roadside soil (Table 3), except for Cu, where a significant negative correlation r = −0.562 was found for CaCl$_2$ extractable Cu. In general, the factors best correlated with shoot metal concentrations were total metal concentrations in soil (Ni: r = 0.750, $p = 0.005$; Zn: r = 0.833, $p = 0.001$), C$_{SOL}$ concentrations (Cu: r = −0.528, $p = 0.009$; Mn: r = −0.544, $p = 0.005$; Ni: r = 0.465, $p = 0.025$; Zn: r = 0.485, $p = 0.016$) and soil pH (Ni: r = 0.632, $p = 0.001$; Zn: r = 0.589, $p = 0.004$).

Table 3. Pearson correlation coefficients (r) between metal concentrations in wheat tissues, soil physiochemical properties and extractable metals measured by different methods in soil. Statistically significant values ($p < 0.05$) are in bold font.

Metal	Variable	C$_{DGT}$	C$_{SOL}$	R	Wheat	Soil	C$_E$	CaCl$_2$	pH	TOC
Cd	CSOL	0.54 *	1							
	R	0.27	−0.27	1						
	Wheat	0.27	0.02	0.37	1					
	Soil	0.41	0.55 *	−0.27	0.48 *	1				
	CE	0.99 ***	0.55 *	0.26	0.25	0.39	1			
	CaCl2	0.22	−0.46	0.48	0.10	0.34	0.21	1		
	pH	0.06	0.26	−0.57 *	0.22	0.57 *	0.04	−0.20	1	
	TOC	0.18	0.23	−0.14	−0.44	0.01	0.20	−0.09	−0.20	1
	TC	−0.23	0.10	−0.25	0.32	0.58 **	−0.25	−0.17	0.63 **	0.08
Cu	CSOL	0.90 ***	1							
	R	0.66 **	0.33	1						
	Wheat	−0.44	−0.53 *	−0.01	1					
	Soil	0.13	0.30	−0.21	0.26	1				
	CE	0.99 ***	−0.91 ***	0.65 **	−0.46	0.13	1			
	CaCl2	0.42	0.57	0.03	−0.58 *	0.25	0.45	1		
	pH	−0.22	0.10	−0.63 **	0.11	0.51 *	−0.22	0.04	1	
	TOC	0.30	0.2	−0.02	−0.23	0.27	0.31	−0.03	−0.20	1
	TC	−0.32	−0.21	−0.55 *	0.11	−0.17	−0.32	0.01	0.36	−0.04
Mn	CSOL	0.13	1							
	R	0.31	0.07	1						
	Wheat	−0.15	−0.54 *	−0.16	1					
	Soil	−0.46	−0.46	0.10	0.27	1				
	CE	0.96 ***	−<0.01	0.26	0.04	−0.50	1			
	CaCl2	0.09	0.27	0.07	0.05	0.81 ***	−0.37	1		
	pH	−0.44	−0.59 **	0.31	0.36	0.54 *	−0.009	−0.20	1	
	TOC	0.43	−0.22	−0.36	0.30	−0.26	0.54 *	−0.009	−0.20	1
	TC	0.17	−0.15	−0.91 ***	0.13	−0.34	0.31	−0.34	−0.25	0.70 **

Table 3. Cont.

Metal	Variable	C_{DGT}	C_{SOL}	R	Wheat	Soil	C_E	$CaCl_2$	pH	TOC
Ni	C_{SOL}	0.56 *	1							
	R	0.04	−0.59 *	1						
	Wheat	0.22	−0.3	0.46	1					
	Soil	0.48 *	0.67 **	−0.2	0.75 ***	1				
	CE	0.99 ***	0.52 *	0.01	0.19	0.42	1			
	CaCl2	0.32	−0.03	0.4 *	−0.33	0.13	0.30	1		
	pH	−0.15	0.45	−0.73 **	0.63 **	0.52 *	−0.19	−0.52 *	1	
	TOC	0.37	0.30	0.10	−0.16	−0.04	0.38	0.26	−0.20	1
	Tc	−0.20	0.50	−0.79 ***	0.20	0.31	−0.19	−0.45	0.66 **	−0.03
Zn	C_{SOL}	0.14	1							
	R	−0.10	−0.97 ***	1						
	Wheat	0.13	0.48 *	−0.45	1					
	Soil	0.43	0.31	−0.32	0.83 ***	1				
	CE	0.99 ***	0.15	−0.10	0.13	0.43	1			
	CaCl2	−0.08	−0.32	0.30	0.07	0.02	−0.07	1		
	pH	−0.48	0.55 *	−0.60 **	0.59 **	0.27	−0.49 *	−0.21	1	
	TOC	0.82 ***	0.42	−0.34	0.35	0.50 *	0.83 ***	−0.09	−0.20	1
	Tc	−0.57 *	0.46	−0.53 *	0.28	−0.06	−0.57 *	0.02	0.81 ***	−0.33

*** Correlation is significant at the 0.001 level, ** Correlation is significant at the 0.01 level, * Correlation is significant at the 0.05 level, Wheat = metal concentrations in wheat shoots, Soil = total metal concentration in soil, R = C_{DGT}/C_{SOL}.

Soil pH also correlated with several other soil metal measurements, in particular with R, C_{SOL} and Tc (Table 3). The dissociation time for Cd (r = 0.630, p = 0.028), Ni (r = 0.661, p = 0.027) and Zn (r = 0.807, p = 0.002) in soil was significantly correlated with soil pH (Table 3), while Tc showed a strong correlation with soil TOC for Mn (r = 0.701).

3.3. Metal Aging and Dissociation Time in Soil

The time for the dissociation of metals from the solid-liquid interface was examined with respect to the effect of the time of soil exposure to vehicular emitted metal contamination (i.e., road age) on metal dissociation in soil (Figure 1). Negative marginally significant (p < 0.1) linear correlations were found between Tc and the following soil metal concentrations: Cd (R^2 = 0.456, p = 0.096), Cu (R^2 = 0.560, p = 0.083) (Figure 1) and Zn (R^2 = 0.578, p = 0.072) (Figure 1). In contrast, the Tc of Mn increased significantly (p < 0.05) with the length of time for contamination (R^2 = 0.617, p = 0.002), and Ni did not show a linear relationship between Tc and road age (R^2 = 0.084, p = 0.874).

Optimised linear models between wheat shoot concentrations, i.e., bioavailable metal concentrations and other soil properties for Mn, Ni and Zn all had R^2 > 0.9, but for Cd and Cu, no relationships were able to be developed (Table 4). Total metal concentrations in soil showed a consistent positive relationship with Ni and Zn metal bioavailability, but a negative relationship for Mn. Overall, the metal bioavailability was best explained by the total metal concentration in soil and the road age, while soil properties such as TOC and soil moisture were also important.

Table 4. Optimised linear models for the relationship between metal concentration in wheat, and soil chemical and physical properties. Models were determined by stepwise regression using list wise forward selection (criteria of α = 0.05 for entry).

Independent Variable	Model	R^2
Metal concentrations in wheat	Cd = No models were able to be developed	-
	Cu = No models were able to be developed	-
	Mn (mg/kg) = 104 + 0.01 TOC (C% dry soil) − 0.9 road age (years) − 0.08 soil Mn (mg/kg)	0.936
	Ni (mg/kg) = −7.78 + 0.73 soil-Ni (mg/kg) + 0.37 soil moisture (%)	0.936
	Zn (mg/kg) = 16.5 + 0.32 soil-Zn (mg/kg) − 0.12 age (years)	0.931

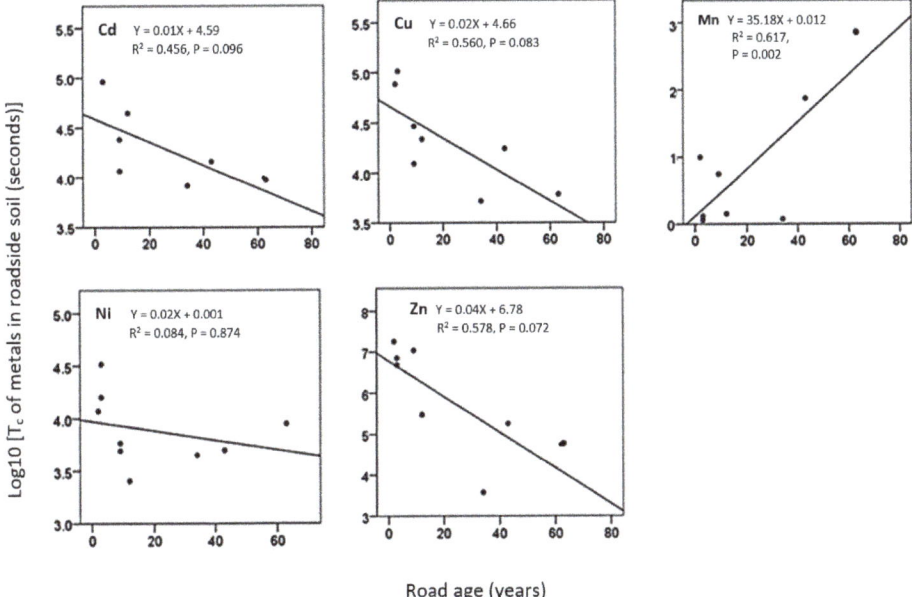

Figure 1. The relationship between roadside soil metal dissociation time Tc (seconds) and the age of the adjacent roads representing soil exposure to vehicular emissions (years). R^2 = linear correlation. Please note that in total 9 data points are included for each element, although some points overlap and are not visible.

4. Discussion

Wheat grown in roadside soils accumulated Cd, Mn, Ni and Zn significantly in shoots compared to the control site grown wheat (Table 2). These findings are an indication of the presence of bioavailable metals in the roadside soils tested. In general, out of all the soil assays tested, correlations between metals in wheat shoots and optimized linear models (Table 4) were the best representation of the bioavailability of metals found in the studied roadside soils. Other studies have shown that the toxicity and mobility of metals in soils are as dependent on the total concentrations of metals in soil [31,32].

The magnitude of the r-values for soil concentration (C_{SOL}) relationships in this current study were weak (Tables 3 and 4) and demonstrates that only a small proportion of the variation in metal bioavailability to wheat could be explained via soil solution concentrations. That is, only about 30% of these metals found in wheat shoot could be explained by C_{SOL} for metal concentrations like Zn and Ni. Thus, the C_{SOL} metal concentrations in soil do not appear to be an appropriate measure of bioavailability in roadside soils where the metals have aged in the soil over a number of years. In comparison, C_{SOL} concentrations have previously been reported as a good measure of bioavailability for Zn in spiked roadside soils [33].

The $CaCl_2$ extraction was also found to be a poor measure for metal bioavailability in roadside soils, with the only significant correlation being for Cu (r = −0.582, p = 0.077). It should be noted that the r value for Cu in the current study, while significant is not strong, suggesting that $CaCl_2$ extraction is a poor predictor of the variability of the metal bioavailability in roadside soil (aged). In comparison, Menzies et al. [34], in a meta-analysis of the literature, reported $CaCl_2$ as a good measure of Cu bioavailability for grasses in soil.

The concentration of metals in roadside soils as measured by DGT did not correlate significantly with metal concentrations in wheat shoots (Table 3). This indicates that, in roadside soils where metals have aged for multiple years, DGT measurements of soils were not a good measure of the bioavailability of metals to wheat shoots. While Zhang et al. [33] found good correlations between DGT measurements and Zn plant uptake, their results were for soils that had been recently spiked with metals collected from roadsides, rather than soils where the metals had been present for extended periods of time.

Even though there has been considerable research into metal bioavailability in soil (including Naidu et al. [2], Smolders et al. [35], Zogaj and Düring [11] and many others) there remains a lack of ability to robustly predict bioavailable metal concentrations in soil. Many soil properties such as pH, organic matter content, clay contents/soil composition, and contaminant physicochemical properties such as oxidation state and aqueous solubility are responsible for controlling the behaviour of metals in soil [4]. These properties can be used to predict the fraction of contaminants that may be bioavailable in a given soil, such that a model which can incorporate many of these individual effects on bioavailability may be more effective. The optimised linear models we developed using soil physiochemical properties, different metal concentrations representing different metal extracting techniques and age of the roadside soil showed very strong relationship for Mn, $R^2 = 0.936$; Ni, $R^2 = 0.936$ and Zn, $R^2 = 0.931$ (Table 4). While all the models developed were dependent on total soil metal concentrations, models developed for Mn and Zn clearly demonstrated the effect of road age, i.e., how long metals had aged in situ on metal bioavailability. Thus, optimised linear models have the potential for use in robustly predicting bioavailable metal concentrations in field soils where the metal concentrations have aged in situ.

The low R (C_{DGT}/C_{SOL}) values determined in this research (Table 3) may be as a result of metals being only slowly resupplied from the soil solid phase [36]. The reasons for the low resupply of metals could include limited metal reservoirs, a slow rate constant of the metal resupply process (Kdl) and/or slow desorption metal kinetics [7]. The results suggest a decrease in the R value in soil metals after the wheat assay compared to prior growing wheat. The low R values may be as a result of a reduction in labile metal concentration in the solution due to plant uptake [37].

The biological impact of metals will depend on the rate of the metal dissociation (desorption) from the soil solid to liquid phase and its subsequent availability to organisms. The Tc marginally significantly ($p \leq 0.1$) decreased with the age of the roads for Cd ($p = 0.096$), Cu ($p = 0.08$) and Zn ($p = 0.072$), while for Mn, Tc significantly increased ($p = 0.002$) with the age of the roads (Figure 1). Therefore, Cd, Cu, Ni and Zn showed a slower release rate from the solid to solution phase compared to Mn in newer roads and a higher rate of release for old roads (Figure 1). This is counterintuitive for all metals except Mn, as it would generally be expected that metals that had aged longer in soils would be more tightly bound and therefore dissociate at lower rates. However, correlation coefficients analysis (Table 3) showed that Tc was positively related to soil pH for Cd ($r = 0.630$, $p = 0.028$), Ni ($r = 0.661$, $p = 0.027$) and Zn ($r = 0.807$, $p = 0.004$) (Table 2). The trend of decreasing soil pH with increasing age of roads (Table 1) is likely to have lowered the Tc for these metals. This is in contrast to Ernstberger et al. [8], who reported no clear trend for Ni between Tc and pH. The decrease of soil pH with the age of the road is possible due to sulphur contained in vehicular emissions, resulting in acid generation in soil [38]. The high Mn concentrations present in the newest roads could have also controlled the labile metal pools of other metals by binding to Mn-oxide. For example the presence of high Mn has been shown to effect the binding ability of Cd with Fe-Mn oxides and organics [39], resulting in lower solubility of the corresponding metal in new roadside soil metals. In addition, Temminghoff et al. [40] observed increases in solubility with increases in the organic carbon content of the soil for Cu and Zn in soil, and a similar scenario may have been operating in the roadside soils measured in the current experiment. Thus, a more alkaline pH, low Mn content and higher TOC in older roadside soils (Table 1) may have caused the higher availability of Cd, Cu and Zn compared to soils besides newer roads. Tc and pH could be used as an important tool in comparing soil metal aging effect in soil solutions, thus metal bioavailability.

5. Conclusions

This study investigated the bioavailability of metals aged in situ in roadside soil. None of the soil assays tested (total metals, DGT, soil solution and $CaCl_2$ extractable metals) were good approaches for measuring the bioavailability of metals aged in roadside soils. Optimised linear models showed good potential for use in predicting bioavailable Mn, Ni and Zn in field contaminated aged soils. Soil concentration (C_{SOL}) relationships in this current study were weak, while $CaCl_2$ showed only one significant correlation with Cu (r = −0.582, p = 0.077). This study provided insights into the understanding of the kinetics of long-term metal aging in field contaminated soil, using roadside soils as the model. The dissociation time, Tc, was used for comparing soil metal aging effects. The results showed that Tc was pH dependent for most metals tested (Cd, Ni and Zn). Soil pH was likely to be the key parameter controlling aging of metals in roadside soil, thus metal bioavailability.

Author Contributions: Conceptualization, S.D.S., T.H., S.M.R., and A.S.B.; methodology, S.D.S., T.H., S.M.R., and A.S.B.; validation, S.D.S., T.H., S.M.R.; A.S.B., and D.V.I.; formal analysis, S.D.S, T.H.S. and D.V.I.; investigation, S.D.S.; resources, S.M.R. and T.H.; data curation, S.D.S. and S.M.R.; writing, original draft preparation, S.D.S.; writing, review and editing, S.D.S., S.M.R., T.H., A.S.B., D.V.I.; visualization, S.D.S.; supervision, S.M.R., T.H., and A.B.; project administration, S.M.R. All authors have read and agreed to the published version of the manuscript.

Funding: S.D.S. would like to thank the School of Civil, Chemical and Environmental Engineering, RMIT University for providing the PhD scholarship and Higher degree by research (HDR) publication fellowship.

Acknowledgments: We would like to thank RMIT University, School of Civil, Chemical and Environmental engineering for SDS's PhD scholarship and Higher Degree by Research Publication Grant (HDRPG), during this work. The authors gratefully acknowledge the assistance provided by Paul Morrison for ICP-MS operation and Sandro Logano, Cameron Crombie and Babu Iyer for technical assistance in the laboratory

Conflicts of Interest: The authors declare no conflict of interest.

References

1. Fairbrother, A.; Wenstel, R.; Sappington, K.; Wood, W. Framework for metals risk assessment. *Ecotoxicol. Environ. Saf.* **2007**, *68*, 145–227. [CrossRef] [PubMed]
2. Naidu, R.; Juhasz, A.; Mallavarapu, M.; Smith, E.; Lombi, E.; Bolan, N.S.; Wong, M.H.; Harmsen, J. Chemical Bioavailability in the Terrestrial Environment-recent advances. *J. Hazard. Mater.* **2013**, *261*, 685–686. [CrossRef] [PubMed]
3. McLaughlin, M.J.; Hamon, R.E.; McLaren, R.G.; Speir, T.W.; Rogers, S.L. A bioavailability-based rationale for controlling metal and metalloid contamination of agricultural land in Australia and New Zealand. *Soil Res.* **2000**, *38*, 1037–1086. [CrossRef]
4. Naidu, R.; Bolan, N.S. Contaminant chemistry in soils: Key concepts and bioavailability. *Dev. Soil Sci.* **2008**, *32*, 9–37.
5. Peijnenburg, W.J.; Zablotskaja, M.; Vijver, M.G. Monitoring metals in terrestrial environments within a bioavailability framework and a focus on soil extraction. *Ecotoxicol. Environ. Saf.* **2007**, *67*, 163–179. [CrossRef] [PubMed]
6. Smart, M.; Zarcinas, G.; Stevens, B.; Barry, D.; Cockley, G.; McLaughlin, T. *CSIRO Land and Water's Methods Manual for ACIAR Project no. LWR1/1998/119*; CSIRO Land and Water: Clayton, Australia, 2004.
7. Harper, M.P.; Davison, W.; Zhang, H.; Tych, W. Kinetics of metal exchange between solids and solutions in sediments and soils interpreted from DGT measured fluxes. *Geochim. Cosmochim. Acta* **1998**, *62*, 2757–2770. [CrossRef]
8. Ernstberger, H.; Davison, W.; Zhang, H.; Tye, A.; Young, S. Measurement and dynamic modeling of trace metal mobilization in soils using DGT and DIFS. *Environ. Sci. Technol.* **2002**, *36*, 349–354. [CrossRef]
9. Ernstberger, H.; Zhang, H.; Tye, A.; Young, S.; Davison, W. Desorption kinetics of Cd, Zn, and Ni measured in soils by DGT. *Environ. Sci. Technol.* **2005**, *39*, 1591–1597. [CrossRef]
10. Ivezić, V.; Almås, Å.R.; Singh, B.R. Predicting the solubility of Cd, Cu, Pb and Zn in uncontaminated Croatian soils under different land uses by applying established regression models. *Geoderma* **2012**, *170*, 89–95. [CrossRef]

11. Zogaj, M.; Düring, R.A. Plant uptake of metals, transfer factors and prediction model for two contaminated regions of Kosovo. *J. Plant Nutr. Soil Sci.* **2016**, *179*, 630–640. [CrossRef]
12. Zeng, S.; Li, J.; Wei, D.; Ma, Y. A new model integrating short-and long-term aging of copper added to soils. *PLoS ONE* **2017**, *12*, e0182944. [CrossRef]
13. Anxiang, L.U.; Zhang, S.; Xiangyang, Q.; Wenyong, W.U.; Honglu, L.I.U. Aging effect on the mobility and bioavailability of copper in soil. *J. Environ. Sci.* **2009**, *21*, 173–178.
14. Tagami, K.; Uchida, S. Aging effect on bioavailability of Mn, Co, Zn and Tc in Japanese agricultural soils under waterlogged conditions. *Geoderma* **1998**, *84*, 3–13. [CrossRef]
15. Buekers, J. Fixation of Cadmium, Copper, Nickel and Zinc in Soil: Kinetics, Mechanisms and Its Effect on Metal Bioavailability. Ph.D. Thesis, Faculteit Bio-ingenieurswetenschappen, Katholieke Universiteit Lueven, Leuven, Belgium, 2007.
16. López-García, P.; Moreira, D. Tracking microbial biodiversity through molecular and genomic ecology. *Res. Microbiol.* **2008**, *159*, 67–73. [CrossRef] [PubMed]
17. Ma, Y.; Lombi, E.; McLaughlin, M.J.; Oliver, I.W.; Nolan, A.L.; Oorts, K.; Smolders, E. Aging of nickel added to soils as predicted by soil pH and time. *Chemosphere* **2013**, *92*, 962–968. [CrossRef] [PubMed]
18. Lock, K.; Janssen, C. Ecotoxicity of chromium (III) to Eisenia fetida, Enchytraeus albidus, and Folsomia candida. *Ecotoxicol. Environ. Saf.* **2002**, *51*, 203–205. [CrossRef] [PubMed]
19. Lock, K.; Janssen, C.R. *Influence of Aging on Metal Availability in Soils, Reviews of Environmental Contamination and Toxicology*; Springer: Berlin, Germany, 2003; pp. 1–21.
20. De Silva, S.; Ball, A.S.; Huynh, T.; Reichman, S.M. Metal accumulation in roadside soil in Melbourne, Australia: Effect of road age, traffic density and vehicular speed. *Environ. Pollut.* **2016**, *208*, 102–109. [CrossRef]
21. De Silva, S.; Ball, A.S.; Indrapala, D.V.; Reichman, S.M. Review of the interactions between vehicular emitted potentially toxic elements, roadside soils, and associated biota. *Chemosphere* **2020**. [CrossRef]
22. Isbell, R. *The Australian Soil Classification*; CSIRO Publishing: Clayton, Australia, 2016; pp. 11–41.
23. Price, R.C.; Gray, C.M.; Frey, F.A. Strontium isotopic and trace element heterogeneity in the plains basalts of the Newer Volcanic Province, Victoria, Australia. *Geochim. Cosmochim. Acta* **1997**, *61*, 171–192. [CrossRef]
24. Werkenthin, M.; Kluge, B.; Wessolek, G. Metals in European roadside soils and soil solution—A review. *Environ. Pollut.* **2014**, *189*, 98–110. [CrossRef]
25. Reichman, S.M. Probing the plant growth-promoting and heavy metal tolerance characteristics of Bradyrhizobium japonicum CB1809. *Eur. J. Soil Boil.* **2014**, *63*, 7–13. [CrossRef]
26. Houba, V.; Temminghoff, E.; Gaikhorst, G.; Van Vark, W. Soil analysis procedures using 0.01 M calcium chloride as extraction reagent. *Commun. Soil Sci. Plant Anal.* **2000**, *31*, 1299–1396. [CrossRef]
27. Rayment, G.E.; Higginson, F.R. *Australian Laboratory Handbook of Soil and Water Chemical Methods*; Inkata Press: Melbourne, Australia, 1992.
28. Zhang, H.; Davison, W.; Knight, B.; McGrath, S. In situ measurements of solution concentrations and fluxes of trace metals in soils using DGT. *Environ. Sci. Technol.* **1998**, *32*, 704–710. [CrossRef]
29. Huynh, T.T.; Laidlaw, W.S.; Singh, B.; Zhang, H.; Baker, A.J. Effect of plants on the bioavailability of metals and other chemical properties of biosolids in a column study. *Int. J. Phytoremediat.* **2012**, *14*, 878–893. [CrossRef]
30. van der Ent, A.; Echevarria, G.; Tibbett, M. Delimiting soil chemistry thresholds for nickel hyperaccumulator plants in Sabah (Malaysia). *Chemoecology* **2016**, *26*, 67–82. [CrossRef]
31. Brümmer, G.W.; Gerth, J.; Herms, U. Heavy metal species, mobility and availability in soils. *J. Plant Nutr. Soil Sci.* **1986**, *149*, 382–398.
32. Rodríguez-Flores, M.; Rodríguez-Castellón, E. Lead and cadmium levels in soil and plants near highways and their correlation with traffic density. *Environ. Pollut. Ser. B Chem. Phys.* **1982**, *4*, 281–290. [CrossRef]
33. Zhang, H.; Lombi, E.; Smolders, E.; McGrath, S. Kinetics of Zn Release in Soils and Prediction of Zn Concentration in Plants Using Diffusive Gradients in Thin Films. *Environ. Sci. Technol.* **2004**, *38*, 3608–3613. [CrossRef]
34. Menzies, N.W.; Donn, M.J.; Kopittke, P.M. Evaluation of extractants for estimation of the phytoavailable trace metals in soils. *Environ. Pollut.* **2007**, *145*, 121–130. [CrossRef]

35. Smolders, E.; Oorts, K.; Van Sprang, P.; Schoeters, I.; Janssen, C.R.; McGrath, S.P.; McLaughlin, M.J. Toxicity of trace metals in soil as affected by soil type and aging after contamination: Using calibrated bioavailability models to set ecological soil standards. *Environ. Toxicol. Chem. Int. J.* **2009**, *28*, 1633–1642. [CrossRef]
36. Yao, Y.; Watanabe, T.; Yano, T.; Iseda, T.; Sakamoto, O.; Iwamoto, M.; Inoue, S. An innovative energy-saving in-flight melting technology and its application to glass production. *Sci. Technol. Adv. Mater.* **2008**, *2*, 025013. [CrossRef] [PubMed]
37. Nolan, A.L.; Zhang, H.; McLaughlin, M.J. Prediction of zinc, cadmium, lead, and copper availability to wheat in contaminated soils using chemical speciation, diffusive gradients in thin films, extraction, and isotopic dilution techniques. *J. Environ. Qual.* **2005**, *34*, 496–507. [CrossRef] [PubMed]
38. Maricq, M.M.; Chase, R.E.; Xu, N.; Podsiadlik, D.H. The effects of the catalytic converter and fuel sulfur level on motor vehicle particulate matter emissions: Gasoline vehicles. *Environ. Sci. Technol.* **2002**, *36*, 276–282. [CrossRef]
39. Maiz, I.; Esnaola, M.V.; Millan, E. Evaluation of heavy metal availability in contaminated soils by a short sequential extraction procedure. *Sci. Total Environ.* **1997**, *206*, 107–115. [CrossRef]
40. Temminghoff, E.J.M.; Van der Zee, S.E.; de Haan, F.A.M. Copper mobility in a copper-contaminated sandy soil as affected by pH and solid and dissolved organic matter. *Environ. Sci. Technol.* **1997**, *31*, 1109–1115. [CrossRef]

Publisher's Note: MDPI stays neutral with regard to jurisdictional claims in published maps and institutional affiliations.

© 2020 by the authors. Licensee MDPI, Basel, Switzerland. This article is an open access article distributed under the terms and conditions of the Creative Commons Attribution (CC BY) license (http://creativecommons.org/licenses/by/4.0/).

Article

Adsorption/Desorption Patterns of Selenium for Acid and Alkaline Soils of Xerothermic Environments

Ioannis Zafeiriou, Dionisios Gasparatos and Ioannis Massas *

Laboratory of Soil Science and Agricultural Chemistry, Agricultural University of Athens, 11855 Athens, Greece; j.zafeiriou@gmail.com (I.Z.); gasparatos@aua.gr (D.G.)
* Correspondence: massas@aua.gr

Received: 22 July 2020; Accepted: 23 September 2020; Published: 24 September 2020

Abstract: Selenium adsorption/desorption behavior was examined for eight Greek top soils with different properties, aiming to describe the geochemistry of the elements in the selected soils in terms of bioavailability and contamination risk by leaching. Four soils were acid and four alkaline, and metal oxides content greatly differed between the two groups of soils. The concentrations of Se(IV) used for the performed adsorption batch experiments ranged from 1 to 50 mg/L, while the soil to solution ratio was 1 g/0.03 L. Acid soils adsorbed significantly higher amounts of the added Se(IV) than alkaline soils. Freundlich and Langmuir equations adequately described the adsorption of Se(IV) in the studied soils, and the parameters of both isotherms significantly correlated with soil properties. In particular, both K_F and q_m values significantly positively correlated with ammonium oxalate extractable Fe and with dithionite extractable Al and Mn, suggesting that amorphous Fe oxides and Al and Mn oxides greatly affect exogenous Se(IV) adsorption in the eight soils. These two parameters were also significantly negatively correlated with soil electrical conductivity (EC) values, indicating that increased soluble salts concentration suppresses Se(IV) adsorption. No significant relation between adsorbed Se(IV) and soil organic content was recorded. A weak salt (0.25 M KCl) was used at the same soil to solution ratio to extract the amount of the adsorbed Se(IV) that is easily exchangeable and thus highly available in the soil ecosystem. A much higher Se(IV) desorption from alkaline soils was observed, pointing to the stronger retention of added Se(IV) by the acid soils. This result implies that in acid soils surface complexes on metal oxides may have been formed restricting Se desorption.

Keywords: selenium; acid soils; alkaline soils; adsorption; desorption; Freundlich; Langmuir; Mediterranean soils

1. Introduction

Selenium (Se) is an essential micronutrient for humans and animals, but can lead to toxicity when taken in excessive amounts. Plants are the main source of dietary Se, but the essentiality of Se for plants is still controversial, although the beneficial effects of low doses of Se on plants have been reported in several studies [1–3]. The concentration of Se in plants is directly related to the concentration and the bioavailability of the element in the soil and the plant species [4]. Selenium reactivity in soils depends not only on its total content but also on its chemical form [5,6]. The mobility and plant-availability of Se in soil is controlled by numerous chemical and biochemical processes, as follows: sorption, desorption, microbial activity, the formation of organic and inorganic complexes, precipitation, and dissolution and methylation to volatile compounds [6,7]. Depending on the oxidation state, Se is present in soil as selenide (Se_2^-), elemental selenium (Se^0), selenite (SeO_3^{2-}), selenate (SeO^{2-}) and organic Se. The main factors controlling Se solubility and availability in soils are considered to be pH, oxidation-reduction potential (Eh), metallic oxy-hydroxides and clays, organic matter, microorganisms, and the presence of competing ions [6,8]. Comprehensive information regarding Se geochemistry and Se behavior in soil–plant systems is included in the extensive reviews of Winkel et al. [6], Etteieb et al. [8] and Schivaon et al. [9]

The total concentration of Se in soils varies spatially, and the average global value is quite low at 0.4 mg kg^{-1}, ranging between 0.01 and 2 mg kg^{-1} [9,10]; soils containing less than 0.5 mg kg^{-1} Se are considered as deficient. In humans, Se deficiency occurs when a dietary intake of Se is <40 μg/day and chronic toxicity is observed above levels of >400 μg/day [11]. WHO has recommended 50–55 μg/day Se in human diet [12–14]. It has been estimated that more than 1 billion people all over the world are suffering Se malnutrition, which makes them susceptible to health problems such as growth retardation, impaired bone metabolism and abnormalities in thyroid function [7,9,12]. Selenium deficiency has been reported in countries such as Canada, China, Scotland, Japan, New Zealand, Spain and USA [6,7,15,16]. Thus, numerous studies have been carried out aiming to enrich agricultural products with Se [17–19], and to examine the behavior of added Se in soils. Greece is also considered as an Se deficient area (daily Se intake <55 μg) [20], and very low selenium concentrations were recorded in Greek agricultural products such as soft and hard wheat, barley, oat, rye and corn [21]. However, published studies reporting on Se concentrations or describing the geochemical behavior of the element in Greek soils are missing from the literature. Considering that Greek soils are Se deficient, it is highly possible that in the future Se addition by fertilization can be proposed in order to enrich edible agricultural products. Thus, the geochemical behavior of Se in soils with different physicochemical properties should be examined to ensure the availability of Se for plant uptake and to restrict Se leaching. It is worth to note that Greek soils can be regarded as representative of Mediterranean soils, and any information on the geochemistry of Se in these soils can be projected and used for soils of similar characteristics formed and developed under comparable environmental conditions.

The purpose of the present study was to obtain data on the behavior of freshly added Se(IV) in acid and alkaline Greek soils with different physicochemical properties, and to evaluate the potential environmental risks arising from Se(IV) application. Thus, a batch experiment was conducted to investigate (a) the adsorption of different Se(IV) concentrations in the selected soils, (b) the desorption patterns of sorbed Se(IV) by using 0.25 M KCl as a desorbing agent, as well as (c) to determine the soil properties that mainly affect the sorption/desorption processes.

2. Materials and Methods

2.1. Soils

Eight composite top soil samples (0–20 cm) representing a range of different physicochemical properties were collected from arable lands of Peloponnese (Greece) and used in this study. The main criterion for the selection of sampling sites was the soil pH. Four of the soils were acid and four alkaline. The samples were transferred in sterile sampling bags to the laboratory, air-dried, crushed, passed through a 2-mm sieve and finally stored again in sterile sampling bags. Particle size distribution was determined by the hydrometer method [22], while pH and EC were measured in a 1:1 (w/v) soil/water ratio [23]. Total carbonates content ($CaCO_3$) was calculated by measuring the evolved CO_2 following HCl dissolution [24]. The Loeppert and Suarez [25] ammonium oxalate method was used in order to determine active carbonate fraction. Available phosphorous (p) was obtained by using the Olsen method [26] and organic carbon (OC) content was determined by the Walkley-Black's protocol [27]. Amorphous and free Fe, Mn and Al oxide contents were calculated by the ammonium oxalate buffer methods [28] and by the sodium–bicarbonate–dithionate (CBD) [29], respectively. Total Se was extracted by aqua regia [30].

2.2. Stock Solutions and Reagents

Stock solutions containing 1, 10, 20, 30, 40 and 50 mg Se(IV) L^{-1} were prepared by diluting the appropriate amount of SeO_2 in deionized water and were stored in airtight sterile glass containers.

The desorbing solution of 0.25 M KCl was prepared by dissolving the proper amount of KCl salt in deionized water. This solution was also stored in airtight sterile glass containers.

2.3. Batch Experiments

For every soil six falcon tubes were used. One gram of soil was introduced to each falcon tube and 30 mL of the appropriate stock solution was added, resulting in a 1:30 w/v soil:solution ratio. Afterwards the falcon tubes were placed in an incubator with an adjusted steady temperature of 22 ± 1 °C and gently shaken at 120 rpm for 24 h on an end-to-end shaker. Then, the falcon tubes were centrifuged for 5 min at 3500 rpm and the supernatants were filtered through a Whatman paper No 42. Absorbed Se(IV) was calculated by the difference between the initial and the equilibrium solutions Se(IV) concentrations. Moreover, since pH plays an important role in Se behavior in the soil environment, the pH values of the initial Se(IV) solutions and of the equilibrium solutions were also recorded.

To desorb adsorbed Se(IV), 30 mL of 0.25 M KCl extractant solution was added in the falcon tubes containing the soil samples. Falcon tubes were placed again in an incubator with an adjusted steady temperature of 22 ± 1 °C, and gently shaken at 120 rpm for 24 h on an end-to-end shaker, centrifuged, and filtered through a Whatman paper No. 42, following the same procedure as described above. Desorbed Se(IV) was determined in the equilibrium solutions at the end of the process.

2.4. Isotherm Equations

Langmuir and Freundlich adsorption isotherms were produced based on the equilibrium adsorption data. However, the Langmuir model assumes that biosorption takes place at specific homogeneous sites on the adsorbent by monolayer coverage, while the Freundlich model is empirical and assumes sorption on a heterogeneous surface.

The linear form of the Langmuir model is [31]

$$\frac{C_e}{q_e} = \frac{1}{q_m}C_e + \frac{1}{b_L q_m} \quad (1)$$

where C_e is the equilibrium concentration of ion in the solution (mg/L), q_e is the amount of ion adsorbed per gram of adsorbent at equilibrium (mg/g), q_m is the monolayer biosorption capacity (mg/g) and b_L is the affinity constant related to the binding strength of adsorption (L/mg). The values of q_m and b_L can be determined from the linear plot of C_e/q_e versus C_e.

The linear form of the Freundlich model is [32]

$$\ln q_e = \ln K_F + \frac{1}{n} \ln C_e \quad (2)$$

where C_e is the equilibrium concentration of ion in the solution (mg/L), q_e is the amount of ion adsorbed per gram of adsorbent at equilibrium (mg/g), K_F is a constant relating to the biosorption capacity (mg/g) (L/mg)$^{1/n}$ and $1/n$ is an empirical parameter relating to the biosorption intensity. The values of K_F and $1/n$ can be determined by plotting $\ln q_e$ versus $\ln C_e$.

2.5. Distribution Coefficient (K_d)

The distribution coefficient (K_d) (L/kg) was calculated according to the following formula:

$$K_d = q_e/C_e \quad (3)$$

where C_e is the equilibrium concentration of ion in the solution (mg/L) and q_e is the amount of ion adsorbed per kg of adsorbent at equilibrium (mg/kg).

2.6. Analytical Determinations

Selenium, iron, manganese and aluminum concentrations were determined by using an atomic absorption spectrophotometry, Varian—spectraAA-300system. For the determination of Se at low concentrations, a Varian model VGA77 hydride generator was used. Available phosphorus concentrations were determined by a Shimadzu UV-1700 spectrophotometer. Every 10 samples a control sample was analyzed, and at the end of the measurements procedure 30% of the samples were reanalyzed to test reproducibility.

2.7. Statistics

Correlation and t-test analysis ($p < 0.05$) were performed using STATISTICA 10 software (StatSoft Inc., Tulsa, 74104 OK, USA).

3. Results

3.1. Soil Properties

The physicochemical properties of the studied soils and the total Se concentrations are summarized in Table 1. Most of the soils are characterized as medium to fine textured with low organic carbon content, as expected for Mediterranean agricultural soils, and with very low total Se concentrations, less than 0.28 mg kg^{-1}, pointing to Se deficiency [33]. Ammonium oxalate and dithionite extractable Fe, Al and Mn are expressed as % oxides content (g 100 g^{-1} soil) and presented as Feo, Alo, Mno and Fed, Ald and Mnd, respectively. Metal oxide concentrations greatly varied, ranging between 0.08 and 0.40% and 0.72 and 6.32% for Feo and Fed, between 0.55 and 1.03% and 0.06 and 0.26% for Alo and Ald and between 0.01 and 0.10% and 0.02 and 0.15% for Mno and Mnd. The pH range of both alkaline and acid soils was very narrow—7.4 to 7.8 for alkaline soils and 5.5 to 6.0 for acid soils. Electrical conductivity values in the alkaline soils were significantly higher than in the acid soils ($p < 0.001, n = 4$), but were not restrictive for the growth of crops. Most soils were marginally to moderately supplied with available phosphorus.

Table 1. Soil physicochemical characteristics.

Soil properties	Alkaline Soils				Acid Soils			
	1	2	3	4	5	6	7	8
Clay (%)	37.6	23.6	17	28.7	24.7	32.4	16.4	30.1
Silt (%)	25.7	32	18	30.3	26.3	24.3	20.3	20.3
Sand (%)	36.7	44.4	65	41	49	43.3	63.3	49.6
Texture	CL	CL	SL	CL	SCL	CL	SL	SCL
pH (1:1)	7.45	7.42	7.44	7.76	5.49	5.88	6.01	5.8
CaCO$_3$ eq. (%)	4.5	4.55	18.7	16.3	<D.L. *	<D.L.	<D.L.	<D.L.
Act. CaCO$_3$ (%)	3.13	2.63	0.5	4.86	<D.L.	<D.L.	<D.L.	<D.L.
EC (µS/cm)	1900	1365	1545	1750	960	625	475	400
Organic Carbon %	1.05	0.95	1.50	0.70	1.55	0.80	0.75	0.85
Fe$_d$ (%)	1.8	0.73	6.32	1.57	2.26	3.22	2.33	1.28
Fe$_o$ (%)	0.2	0.13	0.13	0.08	0.17	0.4	0.31	0.35
Al$_d$ (%)	0.12	0.06	0.06	0.12	0.13	0.26	0.16	0.22
Feo/Fed	0.11	0.18	0.02	0.05	0.08	0.12	0.13	0.27
Al$_o$ (%)	0.9	0.64	0.55	0.66	0.9	1.02	0.46	1.03
Mn$_d$ (%)	0.05	0.04	0.03	0.02	0.09	0.1	0.07	0.15
Mn$_o$ (%)	0.04	0.04	0.02	0.02	0.08	0.1	0.05	0.06
Se total (mg kg^{-1})	0.21	0.28	0.07	0.06	0.16	0.08	0.18	0.05
p Olsen. (mg kg^{-1})	4	18.8	8.8	6.3	11.7	10.6	27.6	6.4

* D.L.: Detection Limit.

3.2. Selenium Adsorption

Acid soils showed a much higher retention of added Se(IV) than alkaline soils, in accordance with many studies [6,34–36]. In particular, Se(IV) adsorption ranged between 8.52 and 234 mg kg^{-1}

(Figure 1a) for alkaline soils, while the corresponding range for acid soils was 19.2–558.9 mg kg^{-1} (Figure 1b).

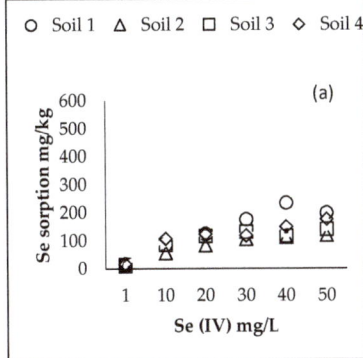
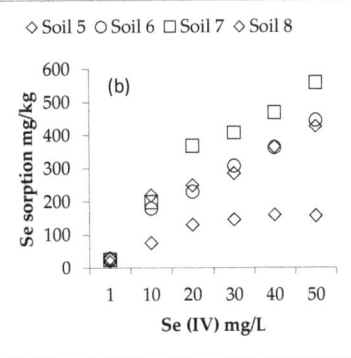

Figure 1. Se(IV) sorption on the studied soils (**a**) alkaline and (**b**) acid. Contact time 24 h, agitation rate 125 rpm, sorbent/solution ratio 1 g/0.03 L, Se(IV) concentrations at start time from 1 to 50 mg/L, temperature 22 °C.

The distribution coefficient (K_d) is a measure of the occupation of available sorption sites in relation to the concentration of the added element. Depending on added Se(IV) concentrations, the Se(IV) K_d values were within the ranges 2.6–36.7 and 3.5–1091.5 L/kg for alkaline and acid soils, respectively. Over the whole range of added Se(IV) concentrations, the K_d values of acid soils were considerably higher than those of the alkaline soils (Figure 2). The observed K_d values for the acid soils were noticeably higher than those reported by Soderlund et al. [36] for selenite adsorption on mineral soils (0.4–240 L/kg), while the highest K_d values are close to those determined by Sheppard et al. [37] for indigenous selenium (800–1500 L/kg). A decreasing trend of K_d values is commonly observed as the concentration of the element in solution increases, indicating that proportionally less of the added element is adsorbed by the soil colloids. Indeed, for all studied soils, K_d decreased as the Se(IV) solution concentration increased (Figure 2), and the higher to lower K_d ratio ranged between 4.6 and 9.1 for alkaline soils, whereas the corresponding range for acid soils was 10–90.2.

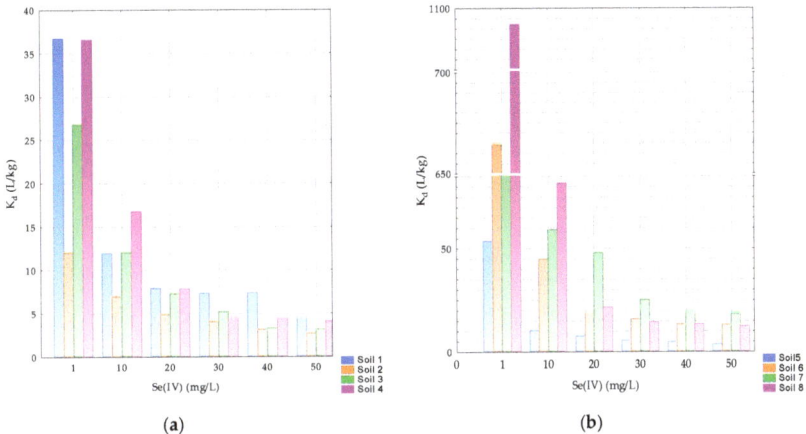

Figure 2. Values of Se(IV) K_d (L/kg) for the studied soils (**a**) alkaline and (**b**) acid. Contact time 24 h, agitation rate 125 rpm, sorbent/solution ratio 1 g/0.03 L, Se(IV) concentrations at start time from 1 to 50 mg/L, temperature 22 °C.

3.3. Selenium Desorption

In the present study, 0.25 M KCl was used to extract adsorbed Se(IV). As is stated by Dhillon and Dhillon [34] and Zhu et al. [38], chloride ion can replace non-specifically adsorbed Se through ion exchange and mass action mechanisms. The desorption pattern was almost identical for all soils, i.e., increasing the initial Se(IV) solution concentration resulted in increasing the Se amounts desorbed from the soils (Figure 3). For all initial Se(IV) concentrations, less Se desorbed from acid soils, a trend more pronounced for initial solution concentrations up to 40 mg Se(IV)/L. Depending on the initial Se(IV) solution concentration, desorbed Se ranged between 2.6 and 117.6 and 0.2 and 84 mg kg^{-1} for alkaline and acid soils respectively (Figure 3).

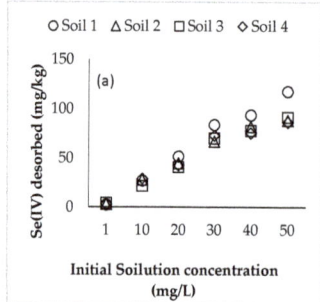
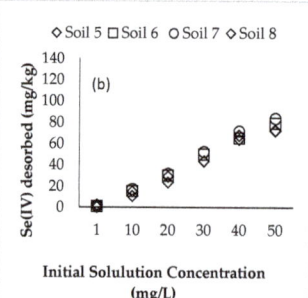

Figure 3. Se(IV) desorption from the studied soils (**a**) alkaline and (**b**) acid. Contact time 24 h, agitation rate 125 rpm, sorbent/solution ratio 1 g/0.03 L, temperature 22 °C.

3.4. Equilibrium Solutions pH

For all soils the acid initial solutions, pH led to acidic equilibrium solutions pH (Figure 4). In particular, the equilibrium solutions' pH values for alkaline soils showed a decrease between one and three units as the concentration of added Se(IV) increased, while for acid soils the corresponding decrease was sharp for a 10 mg/L initial Se(IV) concentration, remaining almost constant thereafter for higher Se(IV) concentrations. Alkaline soils 3 and 4 showed higher resistances to pH changes than alkaline soils 1 and 2, probably due to the higher buffering capacity attributed to the higher carbonates content (Table 1).

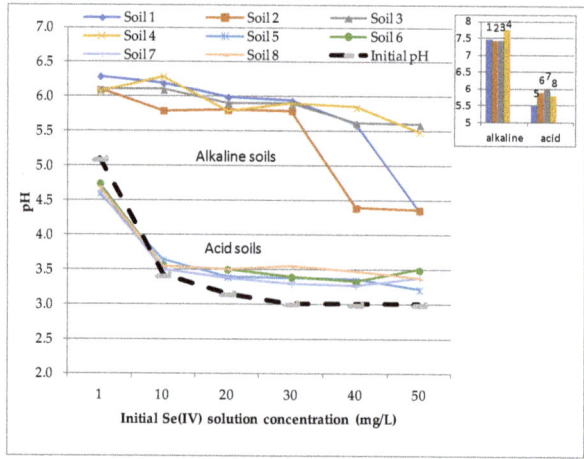

Figure 4. Equilibrium solutions pH values. Dashed line shows the initial solutions pH values. Soil pH values are presented in the incorporated frame.

4. Discussion

4.1. Selenium Adsorption

The experimental data fitted well with Freundlich and Langmuir isotherms, in agreement with Dhillon and Dhillon's results [35] (Table 2). The calculated adsorption maxima (q_m) from the Langmuir isotherm were higher for acid soils, as was in most cases the value of the bonding constant (b_L), indicating the stronger Se(IV) retention by the acid soils.

Table 2. Parameters of the Langmuir and Freundlich models for Se(IV) sorption in the eight soils. Contact time 24 h, agitation rate 125 rpm, sorbent/solution ratio 1 g/0.03 L, Se(IV) concentrations at start time from 1 to 50 mg/L, temperature 22 °C.

Soil	Langmuir Constants				Freundlich Constants			
	q_m (mg/g)	b_L (L/mg)	R^2	p-Value	K_F (mg/g) (L/mg)$^{1/n}$	$1/n$	R^2	p-Value
1	0.26	0.085	0.9	<0.01	4.16	0.578	0.987	<0.001
2	0.15	0.076	0.996	<0.001	2.93	0.648	0.980	<0.001
3	0.15	0.203	0.979	<0.001	3.90	0.514	0.939	<0.01
4	0.18	0.152	0.939	<0.01	4.29	0.492	0.935	<0.01
5	0.18	0.140	0.973	<0.001	4.26	0.458	0.991	<0.001
6	0.46	0.157	0.894	<0.01	7.33	0.394	0.993	<0.001
7	0.61	0.176	0.969	<0.001	6.95	0.571	0.979	<0.001
8	0.42	0.246	0.921	<0.01	7.83	0.355	0.973	<0.001

The parameters of both isotherms, i.e., K_F and $1/n$ from the Freundlich isotherm, and q_m and b_L from the Langmuir isotherm, showed significant correlations with soil constituents. Both K_F and q_m significantly positively correlated with ammonium oxalate extractable Fe and with dithionite extractable Al and Mn ($p < 0.01$, Table 3), underpinning the crucial role of amorphous Fe, Al and Mn oxides in the exogenous Se(IV) behavior of the studied soils. The ability of Fe (especially amorphous), Al and Mn oxides to control Se geochemical behavior has been highlighted in many studies, supporting thus the leading significance of metal oxides in regulating Se mobility in soils [6,10,34,39–41]. K_F and q_m were also significantly negatively correlated with EC ($p < 0.05$, Table 3) and negatively but not significantly with bonding constant (b_L). These relations suggest that an increased soluble salts concentration suppresses both Se(IV) adsorption and strength of Se(IV) retention in soils, and leads to the increased availability of freshly added Se(IV) in the soil environment. This finding is also reported in the review of Natacha et al. [10] and in references therein. Furthermore, the bonding constant (b_L) of the Langmuir isotherm significantly positively correlated with the Feo/Fed values of acid soils and with the eqCaCO$_3$ content of alkaline soils (Table 3), pointing to the fact that in acid soils the fresh Se(IV) retention strength increases when amorphous Fe oxides constitute a larger part of free the Fe oxides, whereas in alkaline soils carbonates may possibly affect Se(IV) sorption. No significant correlation between the organic matter content and the initial or the adsorbed Se(IV) content was observed, a conclusion commonly reached by many researchers. Coppin et al. [42] did not find a direct relation between adsorbed Se and organic material, and suggest that Se may be indirectly sorbed on organic particles by forming associations with surface Fe oxides and clays. Additionally, Soderlund et al. [36] reported the limited importance of organic matter on Se retention compared to Fe and Al phases, even when the latter are incorporated in organic substances. Though clay is considered to affect Se sorption in soils [6,43], no significant correlations emerged between the clay content of the soils and the parameters of the Langmuir and Freundlich isotherms, or the distribution coefficient.

In Table 3, the correlation coefficients for Feo, Ald and Mnd and mean K_d (calculated from K_d values for each initial added Se concentration) relations are presented. The significant correlations between K_d values, ammonium oxalate extractable Fe and dithionite extractable Al and Mn ($p < 0.05$) further support that metal oxides govern Se(IV) sorption in the studied soils. The point of zero charge (PZC) of most Fe-oxides was shown to deviate slightly, ranging usually between pH 7 and 9, while the pH$_{pzc}$ values for various Al oxides reported in the literature vary widely, with a median of 8.6. [44,45]. In the pH range of equilibrium solutions, the Fe and Al oxides are positively charged and can adsorb negatively charged Se species. At low pH values, Mn oxides may have offered additional positively charged sites, since the PZC for most Mn oxides usually occurs at pH < 5 [46,47], leading to the increased adsorption capacity of acid soils. Nakamaru et al. [48], by using ^{75}Se as a tracer, found that the K$_d$ values for selenite adsorption in Japanese soils were highly correlated with the active Al (Alo) and Fe(Feo) content of the soils. Premarantha et al. [49] reached the same conclusion for acid soils from rice-growing areas in Sri Lanka. However, Zhe Li et al. [50] did not observe any significant relation between K$_d$ and Alo and/or Feo concentrations in 18 soils from China, and report only a strong negative correlation between K$_d$ and soil pH values, indicating the stronger adsorption of selenite in acid soils. According to Table 3, the EC of soils was also significantly negatively correlated with mean K$_d$ values ($p < 0.05$). Interestingly, Se availability was not only regulated by the absolute poorly crystallized iron oxides, but also by the relative Feo content in the free iron oxides, as can be deduced from the significant correlation between mean K$_d$ and Feo/Fed values ($p < 0.05$, Table 3). Considering that the Feo/Fed ratio is used as an indicator for soil development, this result leads to the speculation that the stage of soil development can influence added Se(IV) behavior in the soil environment, and ultimately in the food chain. Nevertheless, the soils of the present study may have been formed from different parent materials, and such observations could be case specific, but may also be regarded as an indication for further research.

Table 3. Correlation coefficients, significant at $p < 0.05$ except q_m-Mnd and b_L-EC pairs (in italics) ($n = 8$).

Variables	Fe$_o$	Al$_d$	Mn$_d$	Fe$_o$/Fe$_d$	EC	eqCaCO$_3$%
K$_F$	0.91	0.91	0.80		−0.83	
q_m	0.86	0.75	*0.59*		−0.77	
b_L				0.99 ($n = 4$)	*−0.57*	0.96 ($n = 4$)
Mean K$_d$	0.79	0.85	0.89	0.75		
Mean Se desorption %	−0.86	−0.88	−0.80			

4.2. Selenium Desorption

Selenium desorption, presented as the percentage of the adsorbed Se(IV) concentration found in the equilibrium desorption solutions, increased as the added Se(IV) amounts increased (Figure 5). Much lower Se% desorption from the acid than from the alkaline soils was observed, indicating a stronger retention of fresh Se(IV) by the acid soils. In fact, the mean Se% desorption (the average of Se% values for each initial added Se concentration) from the acid soils was significantly lower than the mean Se% desorption from alkaline soils ($p < 0.01$). Acid soils provided more active sites for the adsorption of negatively charged Se(IV) forms, since when lowering the pH positive charges on soil colloids increase, i.e., there is a higher protonation of surface hydroxyl groups, such as Fe-OH and Al-OH functional groups [36]. However, the stronger retention of Se(IV) by acid soils over the whole concentration range implies the involvement of different sorption mechanisms by the two groups of soils. It is probable that surface complexes may have been formed between Se(IV) species and oxides that lowered the reversibility of sorption process in acid soils. As is shown in Figure 4, for acid soils the pH of equilibrating solutions was very low, supporting the claim that stronger acidic conditions may have occurred close to the surfaces of active soil colloids that could lead to the formation of Se species preferably sorbed on such sites [6,10,40]. On the contrary, Se on the active surfaces of alkaline soils may have been retained mostly as easily exchangeable, thus leading to higher Se desorption by KCl. Numerous studies support the claim that low soil pH favors the higher sorption of Se (independently of Se speciation in equilibrating solutions) [8,51–53] but much less has been done on the evaluation of freshly added Se(IV)'s desorption behavior in acid and alkaline soils. The dominant role of metal oxides in the sorption–desorption behavior of Se(IV) under the conditions of the performed batch experiments is also supported by the significant negative correlations between mean Se% desorption values and oxides concentrations (Table 3).

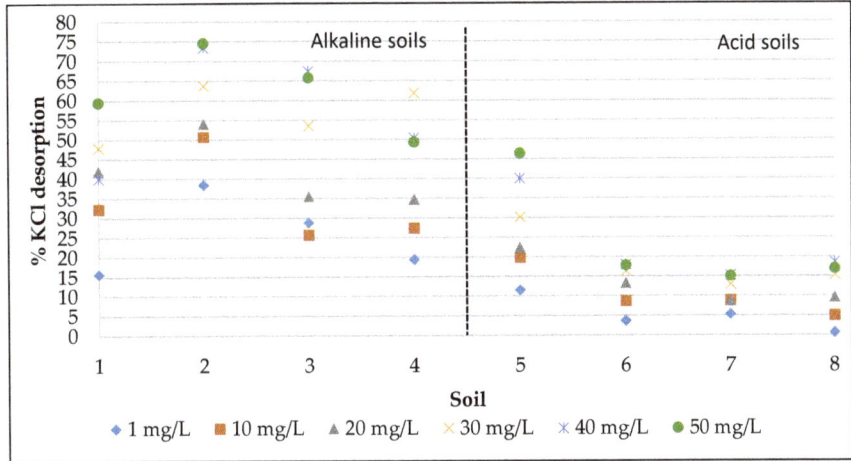

Figure 5. Percentage Se desorption by 0.25 M KCl from alkaline and acid soils. Contact time 24 h, agitation rate 125 rpm, sorbent/solution ratio 1 g/0.03 L, temperature 22 °C.

5. Conclusions

Both the adsorption and desorption processes of freshly added Se(IV) in acid and alkaline soils revealed distinct differences between the two groups of soils. Acid soils adsorbed significantly higher amounts of added Se(IV) than alkaline soils, and alkaline soils desorbed more Se. Fe, Al and Mn oxides, and particularly amorphous Fe oxides content, were the key parameters controlling the sorption/desorption of Se(IV) in the studied soils. Indeed, increased Feo concentration led to higher Se(IV) sorption and to lower Se desorption from the studied soils. Soil pH and the equilibrium solutions' pH strongly influenced both sorption and desorption patterns, providing more positively charged sites on oxides surfaces, leading to higher Se(IV) sorption. Furthermore, metal oxide chemistry at low pH values favored the formation of stronger surface complexes, thus suppressing the Se desorption from acid soils by a weak salt. Overall, the results of this study showed that metal oxides content and pH determine Se geochemistry in soils. Considering that biofortification through plant uptake is also crop/plant-dependent, Se(IV) application in agricultural soils should be site-specific, since a high Se leaching hazard in alkaline soils with low metal oxides concentration may emerge, and low Se availability in acid soils with high metal oxides contents can appear.

Author Contributions: I.Z.: Conceptualization, Methodology, Validation, Formal analysis, Investigation, Data curation, Writing—original draft. D.G.: Methodology, Validation, Resources, Data curation, Writing—review and editing. I.M.: Conceptualization, Methodology, Validation, Formal analysis, Investigation, Resources, Data curation, Writing—original draft, Writing—review and editing, Visualization, Supervision, Project administration. All authors have read and agreed to the published version of the manuscript.

Funding: This research received no external funding.

Conflicts of Interest: The authors declare no conflict of interest.

References

1. Cartes, P.; Jara, A.A.; Pinilla, L.; Rosas, A.; Mora, M.L. Selenium improves the antioxidant ability against aluminium-induced oxidative stress in ryegrass roots. *Ann. Appl. Biol.* **2010**, *156*, 297–307. [CrossRef]
2. Hasanuzzaman, M.; Fujita, M. Selenium Pretreatment Upregulates the Antioxidant Defense and Methylglyoxal Detoxification System and Confers Enhanced Tolerance to Drought Stress in Rapeseed Seedlings. *Biol. Trace Elem. Res.* **2011**, *143*, 1758–1776. [CrossRef] [PubMed]

3. Gupta, S.; Gupta, M. Alleviation of selenium toxicity in *Brassica juncea* L.: Salicylic acid-mediated modulation in toxicity indicators, stress modulators, and sulfur-related gene transcripts. *Protoplasma* **2016**, *253*, 1515–1528. [CrossRef]
4. Dinh, Q.; Wang, M.; Tran, T.; Zhou, F.; Wang, D.; Zhai, H. Bioavailability of selenium in soil-plant system and a regulatory approach. *Crit. Rev. Environ. Sci. Technol.* **2018**, *49*, 443–517. [CrossRef]
5. Finley, J.W.; Davis, C.D.; Feng, Y. Selenium from high selenium broccoli protects rats from colon cancer. *J. Nutr.* **2000**, *130*, 2384–2389. [CrossRef]
6. Winkel, L.; Vriens, B.; Jones, G.; Schneider, L.; Pilon-Smits, E.; Bañuelos, G. Selenium Cycling across Soil-Plant-Atmosphere Interfaces: A Critical Review. *Nutrients* **2015**, *7*, 4199–4239. [CrossRef]
7. Hartikainen, H. Biogeochemistry of selenium and its impact on food chain quality and human health. *J. Trace Elem. Med. Biol.* **2005**, *18*, 309–318. [CrossRef]
8. Etteieb, S.; Magdouli, S.; Zolfaghari, M.; Brar, S. Monitoring and analysis of selenium as an emerging contaminant in mining industry: A critical review. *Sci. Total Environ.* **2020**, *698*, 134339. [CrossRef]
9. Schiavon, M.; Nardi, S.; dalla Vecchia, F.; Ertani, A. Selenium biofortification in the 21st century: Status and challenges for healthy human nutrition. *Plant Soil* **2020**. [CrossRef]
10. Natasha Shahid, M.; Niazi, N.K.; Khalid, S.; Murtaza, B.; Bibi, I.; Rashid, M.I. A critical review of selenium biogeochemical behavior in soil-plant system with an inference to human health. *Environ. Pollut.* **2018**, *234*, 915–934. [CrossRef]
11. Winkel, L.H.; Johnson, C.A.; Lenz, M.; Grundl, T.; Leupin, O.X.; Amini, M.; Charlet, L. Environmental selenium research: From microscopic processes to global understanding. *Environ. Sci. Technol.* **2011**, *46*, 571–579. [CrossRef] [PubMed]
12. World Health Organization (WHO). *Global Health Risks: Mortality and Burden of Disease Attributable to Selected Major Risks*; WHO: Geneva, Switzerland, 2009.
13. Malagoli, M.; Schiavon, M.; Dall'Acqua, S.; Pilon-Smits, E.A.H. Effects of selenium biofortification on crop nutritional quality. *Front. Plant Sci.* **2015**, *6*, 280. [CrossRef] [PubMed]
14. Wu, Z.L.; Bañuelos, G.S.; Lin, Z.Q.; Liu, Y.; Yuan, L.X.; Yin, X.B.; Li, M. Biofortification and phytoremediation of selenium in China. *Front. Plant Sci.* **2015**, *6*, 136. [CrossRef] [PubMed]
15. Shreenath, A.P.; Ameer, M.A.; Dooley, J. *Selenium Deficiency*; StatPearls Publishing: Treasure Island, FL, USA, 2020.
16. Yamada, H.; Kamada, A.; Usuki, M.; Yanai, J. Total selenium content of agricultural soils in Japan. *Soil Sci. Plant Nutr.* **2009**, *55*, 616–622. [CrossRef]
17. Wan, J.; Zhang, M.; Adhikari, B. Advances in selenium-enriched foods: From the farm to the fork. *Trends Food Sci. Technol.* **2018**, *76*, 1–5. [CrossRef]
18. Trolove, S.; Tan, Y.; Morrison, S.; Feng, L.; Eason, J. Development of a method for producing selenium-enriched radish sprouts. *LWT* **2018**, *95*, 187–192. [CrossRef]
19. Longchamp, M.; Castrec-Rouelle, M.; Biron, P.; Bariac, T. Variations in the accumulation, localization and rate of metabolization of selenium in mature Zea mays plants supplied with selenite or selenate. *Food Chem.* **2015**, *182*, 128–135. [CrossRef]
20. Gupta, M.; Gupta, S. An Overview of Selenium Uptake, Metabolism, and Toxicity in Plants. *Front Plant Sci.* **2017**, *7*, 2074. [CrossRef]
21. Bratakos, M.S.; Ioannou, P.V. The regional distribution of selenium in Greeks cereals. *Sci. Total Environ.* **1989**, *84*, 237–247. [CrossRef]
22. Bouyoucos, G.J. A recalibration of the hydrometer method for making mechanical analysis of soils. *Agron. J.* **1951**, *43*, 434–438. [CrossRef]
23. Page, A.L. (Ed.) *Methods of Soil Analysis, Part 2*, 2nd ed.; American Society of Agronomy: Madison, WI, USA, 1982.
24. NF ISO 10693. *Détermination de la Teneuren Carbonate—Méthode Volumétrique*; Qualité des Sols AFNOR: Paris, France, 1995; pp. 177–186.
25. Loeppert, R.H.; Suarez, D.L. Carbonate and gypsum. In *Methods of Soil Analysis, Part 3, Chemical Methods*; Bigham, J.M., Bartels, J.M., Eds.; ASA-SSSA: Madison, WI, USA, 1982; pp. 437–474.
26. Olsen, S.R.; Cole, C.V.; Watanabe, F.S.; Dean, L.A. *Estimation of Available Phosphorus in Soils by Extraction with Sodium Bicarbonate*; US Department of Agriculture: Washington, DC, USA, 1954; Volume 939, pp. 1–19.

27. Nelson, D.W.; Sommers, L.E. Total carbon, organic carbon and organic matter. In *Methods of Soil Analysis, Part 2, Chemical and Microbiological Properties*; Page, A.L., Miller, R.H., Keeney, D.R., Eds.; ASA-SSSA: Madison, WI, USA, 1982.
28. Schwertmann, U.; Taylor, R.M. Iron oxides. In *Minerals in Soil Environments*, 2nd ed.; Dixon, J.B., Weed, S.B., Eds.; 1989; pp. 379–438. Available online: http://www.scielo.br/scielo.php?script=sci_nlinks&ref=000101&pid=S0103-84782013000600009000024&lng=pt (accessed on 17 August 2019).
29. Mehra, O.P.; Jackson, M.L. Iron oxide removal from soils and clay by a dithionite-citrate system buffered with sodium bicarbonate. *Clays Clay Min.* **2013**, *7*, 317–327. [CrossRef]
30. Hagarová, I.; Žemberyová, M.; Bajčan, D. Sequential and single step extraction procedures used for fractionation of selenium in soil samples. *Chem. Pap.* **2005**, *59*, 93–98.
31. Langmuir, I. The adsorption of gases on plane surfaces of glass, mica and platinum. *J. Am. Chem. Soc.* **1918**, *40*, 1362–1403. [CrossRef]
32. Freundlich, H. Über die adsorption in lösungen. *Z. Phys. Chem.* **1906**, *57*, 385–470. [CrossRef]
33. Mirlean, N.; Seus-Arrache, E.R.; Vlasova, O. Selenium deficiency in subtropical littoral pampas: Environmental and dietary aspects. *Environ. Geochem. Health* **2018**, *40*, 543. [CrossRef]
34. Balistrieri, L.S.; Chao, T. Adsorption of selenium by amorphous iron oxyhydroxide and manganese dioxide. *Geochim. Cosmochim. Acta* **1990**, *54*, 739–751. [CrossRef]
35. Dhillon, K.S.; Dhillon, S.K. Adsorption-desorption reactions of selenium in some soils of india. *Geoderma* **1999**, *93*, 19–31. [CrossRef]
36. Söderlund, M.; Virkanen, J.; Holgersson, S.; Lehto, J. Sorption and speciation of selenium in boreal forest soil. *J. Environ. Radioact.* **2016**, *164*, 220–231. [CrossRef]
37. Sheppard, S.C. Robust Prediction of Kd from Soil Properties for Environmental Assessment. *Hum. Ecol. Risk Assess. Int. J.* **2011**, *17*, 263–279. [CrossRef]
38. Zhu, L.; Zhang, L.; Li, J.; Zhang, D.; Chen, L.; Sheng, D.; Yang, S.; Xiao, C.; Wang, J.; Chai, Z.; et al. Selenium sequestration in a cationic layered rare earth hydroxide: A combined batch experiments and EXAFS investigation. *Environ. Sci. Technol.* **2017**, *51*, 8606–8615. [CrossRef]
39. Ligowe, I.; Phiri, F.; Ander, E.; Bailey, E.; Chilimba, A.; Gashu, D.; Joy, E.; Lark, R.; Kabambe, V.; Kalimbira, A.; et al. Selenium deficiency risks in sub-Saharan African food systems and their geospatial linkages. *Proc. Nutr. Soc.* **2020**, 1–11. [CrossRef] [PubMed]
40. Nakamaru, Y.; Altansuvd, J. Speciation and bioavailability of selenium and antimony in non-flooded and wetland soils: A review. *Chemosphere* **2014**, *111*, 366–371. [CrossRef] [PubMed]
41. Rovira, M.; Giménez, J.; Martínez, M.; Martínez-Lladó, X.; de Pablo, J.; Martí, V.; Duro, L. Sorption of selenium(IV) and selenium(VI) onto natural iron oxides: Goethite and hematite. *J. Hazard. Mater.* **2008**, *150*, 279–284. [CrossRef] [PubMed]
42. Coppin, F.; Chabroullet, C.; Martin-Garin, A. Selenite interactions with some particulate organic and mineral fractions isolated from a natural grassland soil. *Eur. J. Soil Sci.* **2009**, *60*, 369–376. [CrossRef]
43. Goldberg, S. Modeling Selenite Adsorption Envelopes on Oxides, Clay Minerals, and Soils using the Triple Layer Model. *Soil Sci. Soc. Am. J.* **2013**, *77*, 64–71. [CrossRef]
44. Kosmulski, M. Evaluation of Points of Zero Charge of Aluminum Oxide Reported in the Literature. *Pr. Nauk. Inst. Gor. Politech. Wroc.* **2001**, *95*, 5–14.
45. Schwertmann, U.; Taylor, R.M. Iron oxides. In *Minerals in Soil Environments*; Dixon, J.B., Weed, S.B., Eds.; Soil Science Society of America: Madison, WI, USA, 1977; pp. 145–179.
46. Miyittah, M.K.; Tsyawo, F.W.; Kumah, K.K.; Stanley, C.D.; Rechcigl, J.E. Suitability of two methods for determination of point of zero charge (PZC) of adsorbents in soils. *Comm. Soil Sci. Plant Anal.* **2016**, *47*, 101–111. [CrossRef]
47. Tan, W.; Lu, S.; Liu, F.; Feng, X.; He, J.; Koopal, L. Determination of the point-of-zero charge of manganese oxides with different methods including an improved salt titration method. *Soil Sci.* **2008**, *173*, 277–286. [CrossRef]
48. Nakamaru, Y.; Tagami, K.; Uchida, S. Distribution coefficient of selenium in Japanese agricultural soils. *Chemosphere* **2005**, *58*, 1347–1354. [CrossRef]
49. Premarathna, H.; McLaughlin, M.; Kirby, J.; Hettiarachchi, G.; Beak, D.; Stacey, S.; Chittleborough, D. Potential Availability of Fertilizer Selenium in Field Capacity and Submerged Soils. *Soil Sci. Soc. Am. J.* **2010**, *74*, 1589–1596. [CrossRef]

50. Liang, D.; Li, Z. Response to the comment by Sabine Goldberg on: Selenite adsorption and desorption in main Chinese soils with their characteristics and physicochemical properties. J. Soils Sediments 2015, 15, 1150–1158, doi:10.1007/s11368-015-1085-7. *J. Soils Sediments* **2016**, *16*, 325. [CrossRef]
51. Antoniadis, V.; Levizou, E.; Shaheen, S.; Ok, Y.; Sebastian, A.; Baum, C. Trace elements in the soil-plant interface: Phytoavailability, translocation, and phytoremediation–A review. *Earth-Sci. Rev.* **2017**, *171*, 621–645. [CrossRef]
52. Loffredo, N.; Mounier, S.; Thiry, Y.; Coppin, F. Sorption of selenate on soils and pure phases: Kinetic parameters and stabilisation. *J. Environ. Radioact.* **2011**, *102*, 843–851. [CrossRef] [PubMed]
53. Garcia-Sanchez, L.; Loffredo, N.; Mounier, S.; Martin-Garin, A.; Coppin, F. Kinetics of selenate sorption in soil as influenced by biotic and abiotic conditions: A stirred flow-through reactor study. *J. Environ. Radioact.* **2014**, *138*, 38–49. [CrossRef] [PubMed]

© 2020 by the authors. Licensee MDPI, Basel, Switzerland. This article is an open access article distributed under the terms and conditions of the Creative Commons Attribution (CC BY) license (http://creativecommons.org/licenses/by/4.0/).

Article

Short- and Long-Term Biochar Cadmium and Lead Immobilization Mechanisms

Liqiang Cui [1], Lianqing Li [2,*], Rongjun Bian [2], Jinlong Yan [1,*], Guixiang Quan [1], Yuming Liu [3], James A. Ippolito [4] and Hui Wang [1]

1. School of Environmental Science and Engineering, Yancheng Institute of Technology, No. 211 Jianjun Road, Yancheng 224003, China; lqcui8@hotmail.com (L.C.); qgx@ycit.cn (G.Q.); whsl@ycit.cn (H.W.)
2. Institute of Resources, Ecosystem and Environment of Agriculture, Nanjing Agricultural University, 1 Weigang, Nanjing 210095, China; brjun@njau.edu.cn
3. Tinghu Station of Agricultural Environment Monitoring, No. 59 Xiwang road, Yancheng 224003, China; liuyuming102986@126.com
4. Department of Soil and Crop Sciences, Colorado State University, Fort Collins, CO 80523, USA; Jim.Ippolito@ColoState.edu
* Correspondence: lqli@njau.edu.cn (L.L.); yjlt@ycit.cn (J.Y.)

Received: 10 June 2020; Accepted: 12 July 2020; Published: 16 July 2020

Abstract: The mechanisms of soil Cd and Pb alterations and distribution following biochar (BC; 0 to 40 t ha^{-1}) amendments applied (in either 2009 [long-term] or in 2016 [short-term]) to a contaminated rice paddy soil, and subsequent plant Cd and Pb tissue distribution over time was investigated. Water-soluble Cd and Pb concentrations decreased by 6.7–76.0% (short-term) and 10.3–88.1% (long-term) with biochar application compared to the control. The soil exchangeable metal fractions (i.e., considered more available) decreased, and the residual metal fractions (i.e., considered less available) increased with short- and long-term biochar amendments, the latter likely a function of biochar increasing pH and forcing Cd and Pb to form crystal mineral lattice associations. Biochar application reduced Cd (16.1–84.1%) and Pb (4.1–40.0%) transfer from root to rice grain, with rice Cd and Pb concentrations lowered to nearly Chinese national food safety standards. Concomitantly, soil organic matter (SOM), pH and soil water content increased by 3.9–49.3%, 0.05–0.35 pH units, and 3.8–77.4%, respectively, with increasing biochar application rate. Following biochar applications, soil microbial diversity (Shannon index) also increased (0.8–46.2%) and soil enzymatic activities were enhanced. Biochar appears to play a pivotal role in forcing Cd and Pb sequestration in contaminated paddy soils, reducing heavy metal transfer to rice grain, and potentially leading to reduced heavy metal consumption by humans.

Keywords: biochar; cadmium; lead; contaminated paddy soil; short- and long-term mechanisms

1. Introduction

Soil heavy metal pollution is a worldwide problem, with accumulation leading to toxicity and environmental persistence; this is especially true for cadmium (Cd) and lead (Pb), both of which pose serious human and ecological health threats [1,2]. In China, over 16% of all agricultural soils are contaminated with heavy metals, with soil Cd and Pb concentrations exceeding China's environmental standards in 7% and 1.5% of all arable lands (9.8 × 10^4 and 2.1 × 10^4 km^2, respectively [3]. It is obvious that in order to protect human health and the environment, reducing heavy metal bioavailability is of paramount importance.

Reducing heavy metal bioavailability has followed several pathways, including amendment additions for heavy metal stabilization [4]. Heavy metal stabilization is typically dominated by chemisorption mechanisms, with other sorption mechanisms such as ion exchange, electrostatic attraction and complexation playing roles [1]. Stabilization/immobilization has been considered practical with respect to short-term effectiveness when using different stabilizers, yet longer-term effectiveness does require more research, especially for the biochar [5]. For example, Senneca et al., found that a cement-based stabilization/solidification treatment positively affected a chromium (Cr) contaminated soil, with the treatment increasing the amount of Cr in the soil residual fraction and thus reducing its mobility [6]. Cui et al., found that the biochar could stabilize Cd and Pb in the paddy soil and reduce the rice uptake in short term [7]. Finding a heavy metal stabilizing material that is effective both in the short- and long-term may be as simple as utilizing biochar [8].

Biochars are created by pyrolyzing various carbonaceous materials, such as agricultural crops wastes, at moderate temperature under anoxic conditions [9]. Biochar is widely accepted as an effective agent for reducing heavy metal bioavailability via biochar organic-heavy metal complexation and biochar oxide, hydroxide, and carbonate phase-heavy metal precipitation [10]. As examples, Khan et al., showed that corn straw biochar could effectively sorb Cd via chemisorption, electrostatic interactions, and inner-sphere complexation reactions [11]. Golden shower tree (*Cassia fistula*) biochar has been shown to sorb and remove up to 303.5 mg Cu g^{-1} from wastewater [12]. In addition, barley grass biochar has been proven to sorb ~90 to 95% of soil borne Cu and Pb from a contaminated soil, leading to enhanced plant growth [13].

Increasing heavy metal sorption via biochar use can lead to alterations in heavy metal phases present in contaminated soils. Essentially, relatively high bioavailable heavy metal concentrations can be altered to less bioavailable forms via biochar application to metal-contaminated soils. Qin et al., showed that the addition of pig manure biochar to contaminated soil sorbed both Cd and Pb, reducing their leaching losses (i.e., bioavailable forms) by 38% and 71%, respectively, as compared to a control [14]. Ippolito et al., observed up to an 88% and 100% decrease in bioavailable Cd and Pb, respectively, with the use of either lodgepole pine or tamarisk biochar in metal contaminated soils; decreases were driven by precipitation reactions [15]. Cadmium sorption onto wheat straw biochar was driven by precipitation reactions [e.g., $Cd(OH)_2$ and $CdCO_3$] and interaction with carbonyl and carboxyl groups, leading to reduced bioavailable soil Cd concentrations [16]. Water hyacinth biochar has been shown to decrease rice paddy soil exchangeable Cd content by ~25%, while increasing Cd in less-bioavailable forms such as those associated with carbonate- and Fe/Mn oxide phases [17].

Altering heavy metal phases present, in favor of lower bioavailability, has been linked to alterations in plant and human metal uptake. For example, corn straw biochar has been shown to significantly reduce the proportion of Cd in the soil exchangeable and carbonate phases (i.e., relatively highly bioavailable) and increase the proportion of Cd in the residual fraction (i.e., highly unavailable), leading to a decrease in plant and human bioavailability [18]. Bian et al., reported that wheat straw biochar reduced Cd and Pb bioavailability by ~60% via sorption onto biochar (hydr)oxide phases present (i.e., unavailable), with Cd and Pb rice uptake reduced by between 27% and 69%, suggesting lower heavy metal consumption by humans [19].

The above studies indicate that biochars may be used to not only sorb, but to alter heavy metals forms to those less bioavailable. However, the additional benefit of biochar land application lies in the fact that biochars can also positively alter soil physicochemical attributes. Zhang et al., showed that rice straw biochar sorbed and immobilized heavy metals, while improving soil water and nutrient dynamics [20]. Bamboo biochar has been shown to increase acidic soil pH, available K, Fe, Mg, and Mn content, and SOM, while decreasing bioavailable heavy metal mobility via surface adsorption and precipitation reactions [21]. Cui et al., reported that biochar increased soil pH and SOM, and significantly reduced metal bioavailability, leading to a decrease in rice and wheat grain metal concentrations over the short-term (i.e., 2 years) [7,22]. Cui et al., showed a similar long-term (i.e., 5 years) trend with wheat straw biochar-Cd/Pb sorption in an amended paddy soil [23].

Based on the information presented above, continued research is needed to verify biochar alterations to soil properties, heavy metal bioavailability, and plant metal uptake over short- and longer-term timeframes. Desperately required is a focus on in-field research. Thus, the objective of the present study was to evaluate the mechanisms by which biochar alters soil physicochemical as well as biological properties, soil Cd and Pb bioavailability, and the corresponding metal uptake in rice from a contaminated paddy soil under the short- and long-term. Our hypothesis was that biochar would bind Cd and Pb and reduce the soil bioavailable fraction/increase the recalcitrant fraction, leading to less Cd and Pb uptake by rice over both a relatively short and longer timeframe. This hypothesis should be supported by correlations between chemical, physical, and biological changes in soil due to biochar application.

2. Materials and Methods

2.1. Site Description

The experiment was conducted in a paddy field (31°24.434′ N and 119°41.605′ E), where atmospheric fallout and effluent discharges from a local Pb smelter have contaminated the site since at least the 1970s. Cadmium and Pb are the primary metal contaminants in this area. The paddy soil was characterized as a Ferric-accumulic Stagnic Anthrosol [24].

2.2. Experimental Design

The long- and short-term biochar experiments commencement in May 2009 and May 2016, respectively, within a continuous rice rotation. The long- and short-term plots were adjacent to one another in the field, using the same experimental design for both experiments; within the text below, the short and long-term experiments are designated as "S" and "L". Individual 4 × 5 m plots were laid out in a randomized complete block design with three replicates per treatment.

Biochar was created from wheat straw via pyrolysis at ~450 °C at the Sanli New Energy Company, Henan Province, China. Treatments included four biochar application rates of 0(C0), 10 (C1), 20 (C2) and 40 (C3) t ha^{-1}, which were surface-applied, raked to relative uniformity, and then fully incorporated into the soil to approximately a 15 cm depth by plowing. Biochar and background soil were collected, returned to the lab, ground to pass a 2-mm sieve, with basic biochar and soil properties determined using methods outlined by Lu [25]; data are presented in Table S1.

2.3. Sample Collection and Analysis

2.3.1. Donnan Membrane Technique Setup

The Donnan membrane technique (DMT), which identifies water-soluble Cd and Pb throughout the rice growing season, was used in-field for every month from August to November 2016, 2017, and 2018. This in-situ technique utilizes an acceptor cell (i.e., 10 mL of 0.01 mol L^{-1} CaCl$_2$, with Ca moving out of the cell when heavy metal cations move into the cell), separated by a positively charged membrane (BDH, No. 55165 2U) held by two O-rings [26] (Figure S1). The BDH membrane has a matrix of polystyrene/divinylbenzene, with sulphonic acid groups which are fully deprotonated above pH = 2. The solution was changed every month over the study period. Solution Cd and Pb concentrations were determined by graphite furnace atomic absorption spectrometry (GFAAS, Zeenit 700p, Jena, Germany).

2.3.2. The Rice Plant Collection

During harvest, three whole rice plants were randomly collected from each plot by gently pulling the entire plant out of the ground. The samples were washed with tap water and deionized water to remove soil, and then separated into roots, shoots, and grain. The plant samples were dried at 105 °C for 30 min and then at 60 °C until dry, then crushed with a pulverizer and stored in air-tight polyethylene bags. A 0.50 g plant sample was placed in a 100 mL beaker and predigested overnight in a 10 mL solution of concentrated HNO$_3$ and HClO$_4$ (4:1, *v:v*) at room temperature. The following day, the beakers were placed on an electric heating plate with the temperature raised from 100 to

200 °C over 30 min, then the temperature was increased to 250 °C until the solution changed colorless, at ~2 mL of solution remaining. The solutions were then removed, cooled, brought to a 25-mL final volume, and filtered through a 0.45-μm membrane filter. The Cd and Pb concentration in the digestate were determined with GFAAS.

2.3.3. The Soil Samples Collection

Three soil cores (0 to 15 cm depth) were collected from each plot following rice harvest in November 2016, 2017 and 2018. Plant debris was removed and the soil separated into two parts: (1) air-dried at room temperature and then ground to pass a 2-mm sieve; (2) stored at 4 °C for subsequent microbial activity measurements. Soil pH, soil water and soil organic matter (SOM) were analyzed using methods described by Lu [25]. Briefly, soil pH was measured using a glass electrode with a soil-to-water ratio of 1:2.5. Soil water content was determined gravimetrically by weight difference before and after at least six hours of oven drying at 105 °C. The SOM was determined using the dichromate oxidation method. A subsample of the air-dried, ground soil was further ground to pass a 0.15-mm sieve and used to determine total Cd and Pb concentrations and sequentially extractable heavy metals fractions as described below [1,7].

2.3.4. The Soil Heavy Metals Fractions Detection

Soil heavy metals from the 2016, 2017, and 2018 samples were sequentially extracted using the modified four-stage procedure recommended by the European Community Bureau of Reference (BCR), as described by Ure et al. and briefly described here. Exchangeable fraction (B1; includes soluble, exchange site- and carbonate-bound) [27]: 40 mL of 0.11 M acetic acid was added to 1 g of soil in a 50 mL centrifuge tube and shaken for 16 h at room temperature, followed by centrifugation and liquid decantation. Iron and manganese oxyhydroxides fraction (B2): 40 mL of a freshly prepared hydroxyl ammonium chloride was added to the residue from the previous step, shaken for 16 h at room temperature, followed by centrifugation and decantation. Organic fraction (B3): the residue from the previous step was treated twice with 10 mL of 8.8 M hydrogen peroxide. The digestion was allowed to stand for 1 h with occasional manual shaking, followed by digestion at 85 ± 2 °C until the volume was reduced to 2–3 mL. Samples were allowed to cool, and then 50 mL of 1.0 M ammonium acetate was added to the mixture and shaken for 16 h at room temperature, followed by centrifugation and decantation. Residual fraction (B4; represents metals associated with crystalline mineral phases): residue from the previous step was allowed to air-dry and then digested in a 50 mL digestion tube by first adding 1 mL of deionized water to make a slurry, followed by an aqua regia addition (7 mL HCl and 2 mL HNO_3). The mixtures were allowed to predigest overnight at room temperature and then digested at 105 °C for 2 h the next day. Then, the mixtures were brought to a 50 mL final volume. The final liquids from all four steps were passed through a 0.45-micron membrane filter prior to Cd and Pb analysis via GFAAS (GFAAS, Zeenit 700 p; Analytik Jena AG, Jena, Germany).

2.3.5. The Soil Enzymatic Activity Detection

Soil enzymatic activity, microbial community composition, and bacterial abundance were analyzed on all refrigerated soil samples. Soil alkaline phosphatase activity was measured using the disodium phenyl phosphate method [28]. Soil sucrase activity was measured using the 3,5-dinitrosalicylic acid method [1]. Soil urease and dehydrogenase activities were determined using phenol-sodium hypochlorite and triphenyl tetrazolium chloride methods, respectively [29]. Microbial community composition was determined using a Gene Amp PCR-System 9700 (Applied Biosystems, Foster City, CA, USA). Briefly, the total DNA was extracted from 0.5 g of soil using a FastDNA Spin Kit for Soil and the FastPrep Instrument (MP Biomedicals, Santa Ana, CA, USA). The DNA quality was assessed on 1% agarose gel, while the quantity of DNA was determined using a Nanodrop-2000 spectrophotometer (Nanodrop Technologies Inc., Wilmington, DE, USA) [30]. Bacterial abundance was quantified following the polymerase chain reaction (PCR) method targeting 515F and 907R (V4-V5 region) primer pairs of 16S rDNA. For microbial community analysis, PCR tests were conducted for each DNA sample, and pooled and purified using a QIAquick Gel Extraction Kit (Qiagen, Chatsworth, CA, USA). Approximately equimolar amounts of the PCR products of each sample were combined prior to amplicon sequencing using an Illumina Miseq platform at Shanghai Genesky Biotechnologies (Shanghai, China).

2.4. Statistical Analysis

All data were expressed as means ± one standard deviation of the mean. Differences between the treatments were examined using a two-way analysis of variance (ANOVA), with statistical differences considered when $p < 0.05$. All statistical analyses were carried out using SPSS, version 20.0 (SPSS Institute, Chicago, IL, USA). Biochar and soil basic properties, plant heavy metals, microbial diversity, heavy metals present in various soil fractions, and soil enzyme activities were analyzed using principle component analysis (PCA, using SPSS), and PCA was also used to determine correlation coefficients (r) between all of these factors.

The Shannon index (H) of the operational taxonomic units (OTU) for soil microbial community diversity was calculated as:

$$H = \sum_{i=1}^{S_{obs}} \frac{n_i}{N} ln \frac{n_i}{N} \qquad (1)$$

where S_{obs} was the OTU detection number, n_i was the number of the ith OTU, and N was the sum of all sequence numbers. A bioconcentration factor (BCF) was determined as the above-ground Cd or Pb concentration/soil Cd or Pb concentration. A translocation factor (TF) was determined by the above-ground plant Cd or Pb concentration/root Cd or Pb concentration.

3. Results and Discussion

3.1. Biochar Effects on Cd and Pb in the Soil Water Soluble Phase and Various Pools

In-situ dissolved soil Cd and Pb concentrations may be considered bioavailable [31]. In-field, soluble soil Cd and Pb data, collected via the DMT, are presented in Figure S2. More often than not, increasing biochar application rate decreased paddy field soil water-soluble Cd and Pb concentrations over both the short- and long-term. Long-term Cd concentrations significantly decreased by 24.5–52.2% (2016), 32.1–73.2% (2017) and 37.1–80.8% (2018), while Pb concentrations decreased by 32.1–88.1% (2016), 15.7–82.4% (2017) and 10.3–46.2% (2018) (Figure S2A,C) with biochar applications compared to the control. Short-term Cd and Pb concentration were significantly decreased by 19.8–46.5% (2016), 14.5–67.4% (2017), 30.0–76.0% (2018) and 6.7–63.3% (2016), 17.6–65.5% (2017), 14.3–44.6% (2018) (Figure S2B,D), respectively, with biochar applications compared to the control. Greater short- and long-term Cd and Pb decreases were associated with greater biochar application rates. When soluble Cd increases were observed, they may have been related to increases in water soluble organic phases present that chelated Cd, as suggested by Fan et al. [32]. However, similar to most of our findings, Xu et al., also found that water-soluble Cd and Pb concentrations decreased by 59% and 13% with macadamia nutshell biochar amendment in a lab incubation study [33]. Comparable Cd and Pb results were found by Wang et al., when using rice straw biochar in a pot experiment [34].

Results of the BCR sequential extraction on soil Cd and Pb fractions are presented in Figure 1. The exchangeable Cd and Pb fractions (B1) were approximately 40% of the total Cd and Pb extracted, yet tended to decrease with increasing biochar application rate. When exchangeable Cd and Pb content decreased, it appears that increases were associated with the residual fraction (B4), indicating a reduction in Cd and Pb bioavailability and lower potential ecological risk associated with biochar amendment, as suggested by others [35]. In support of these findings, Chen et al., showed that biochar reduced Cd bioavailability by transforming the exchangeable fraction into the residual fraction, associating these changes to increases in soil pH [36]. Wang et al. [37] and Liu et al. [38] both observed similar responses when using wheat straw or coconut shell biochar, respectively, to reduce Cd bioavailability. Sludge-based biochars have also been shown to decrease Pb and Cd bioavailability from 55.9% to 4.9% for Pb, and from 78.2% to 12.5% for Cd [39].

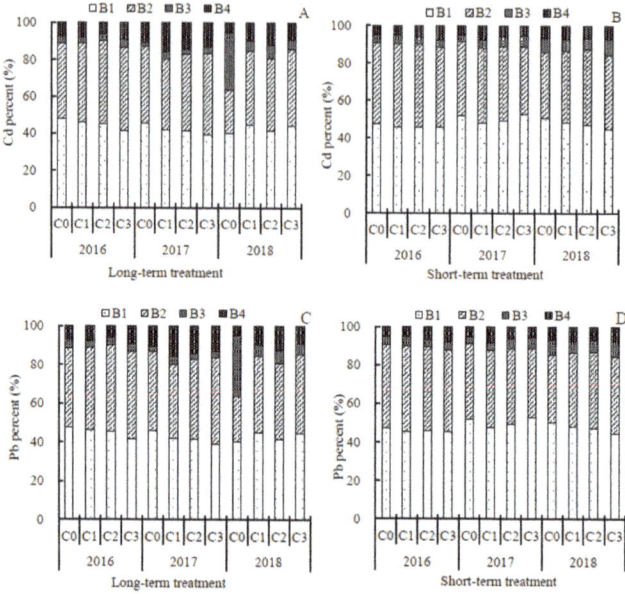

Figure 1. The effect of long- and short-term, increasing biochar application rates [0 (C0), 10 (C1), 20 (C2) and 40 (C3) t ha^{-1}] on Cd and Pb fractions based on a European Community Bureau of Reference (BCR) sequential extraction ((**A**): Cd long-term; (**B**): Cd short-term; (**C**): Pb long-term; (**D**): Pb short-term). B1 = Exchangeable fraction; B2= Iron and manganese oxyhydroxides fraction; B3 = Organic fraction; B4 = Residual fraction.

3.2. Biochar Effects on SOM, pH and Water Content

Changes in soil properties, as a function of biochar application rate in the long-and short-term experiments, are shown in Figure 2. The SOM content significantly increased by 8.1–38.5% (long-term) and 3.9–49.3% (short-term), related to increasing biochar application rate (Figure 2A). The long- and short-term SOM changes remained statistically unchanged over time. Others have also noted similar SOM responses due to biochar application. For example, orchard prunings biochar (10%, *v:v*) reduced exchangeable metal concentrations in soil, which in part was attributed to increasing SOM [40].

Soil water contents were also significantly affected by increasing biochar application rates (Figure 2B). Soil water content increased by 3.8–41.6% (long-term, except 2016) and 4.5–77.4% (short-term), suggesting that biochar application may have improved soil physical properties. Positive changes in soil water content via biochar application have been observed by others [41–44].

Increasing biochar application rates also significantly increased soil pH, with soil pH having been shown to directly influence heavy metal fractions [45]. Soil pH significantly increased by 0.05–0.31 pH units (long-term) and 0.07–0.35 pH units (short-term), with greater changes associated with greater biochar application rates. The short-term usually had a greater effect on increasing soil pH compared to the long-term under the same treatment (Figure 2C). Increasing soil pH has been found to be a key factor for reducing heavy metal bioavailability [15,46].

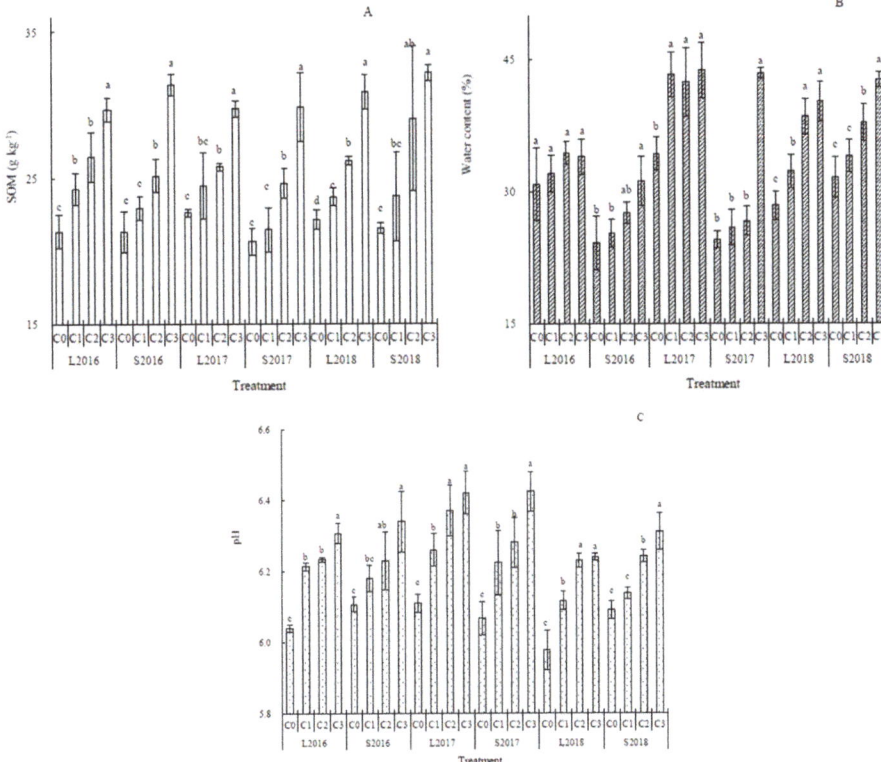

Figure 2. The effect of long- (L) and short-term (S), increasing biochar application rates [0(C0), 10 (C1), 20 (C2) and 40 (C3) t ha^{-1}] on soil (**A**) SOM, (**B**) water content, and (**C**) pH at rice harvest. Different lower-case letters above error bars indicate significant differences between the biochar treatments in either the short- or long-term for a given year [$p < 0.05$, Least significant difference (LSD) post-hoc test]. Error bars represent standard deviation of the mean ($n = 3$).

3.3. Biochar Effects on Cd and Pb Transfer in Rice

The Cd and Pb concentrations in rice, as a function of biochar application rate in the long- and short-term, are shown in Figure 3. Following either short- or long-term biochar application at 40 t ha^{-1}, rice Cd and Pb concentrations decreased by 30.7% and 45.2% (2016), 45.0% and 40.0% (2017), and 84.1% and 28.1% (2018), respectively, as compared to the control; lesser reductions were observed with lower biochar application rates. The rice husk, stem and root Cd and Pb concentrations followed similar trends. Biochar significantly reduced Cd and Pb transfer from roots to stems to grain; root Cd and Pb concentrations were over 40 mg kg^{-1} and 300 mg kg^{-1}, respectively, yet rice grain Cd and Pb concentrations were approaching or met national food safety level ($\leq$0.2 mg kg^{-1}) [47]. Other studies have shown that brinjal (i.e., eggplant) fruit Cd concentrations can be significantly decreased (up to 86.6%) using miscanthus biochar (1.5%, *w:w*) [48].

The BCF and TF were used to assess Cd and Pb transfer from the soil to the above-ground plant, and from plant roots to above-ground tissues, respectively. The BCF decreased by up to 77.1% (Cd) and 33.2% (Pb) in the long-term, and by up to 45.7% (Cd) and 42.3% (Pb) in the short-term. The TF also decreased by up to 53.2% (Cd) and 17.1% (Pb) in the long-term, and by up to 21.9% (Cd) and 23.2% (Pb) in the short-term (Table S2). The trend of decreasing BCF and TF with increasing biochar application rate supports our previous findings that biochar can reduce Cd and Pb bioavailability, suggesting that biochar may play a role in decreasing Cd and Pb transfer in the food chain [49]. Similar to our study,

Mujtaba Munir et al., found that the bamboo biochar treatments reduced TF Cd and Pb by 49.6–61.0% and 61.0–70.7%, respectively, as compared to a control [21]; findings suggested that biochar effectively reduced bioavailable metal phases, leading to reduced metal translocation within plants.

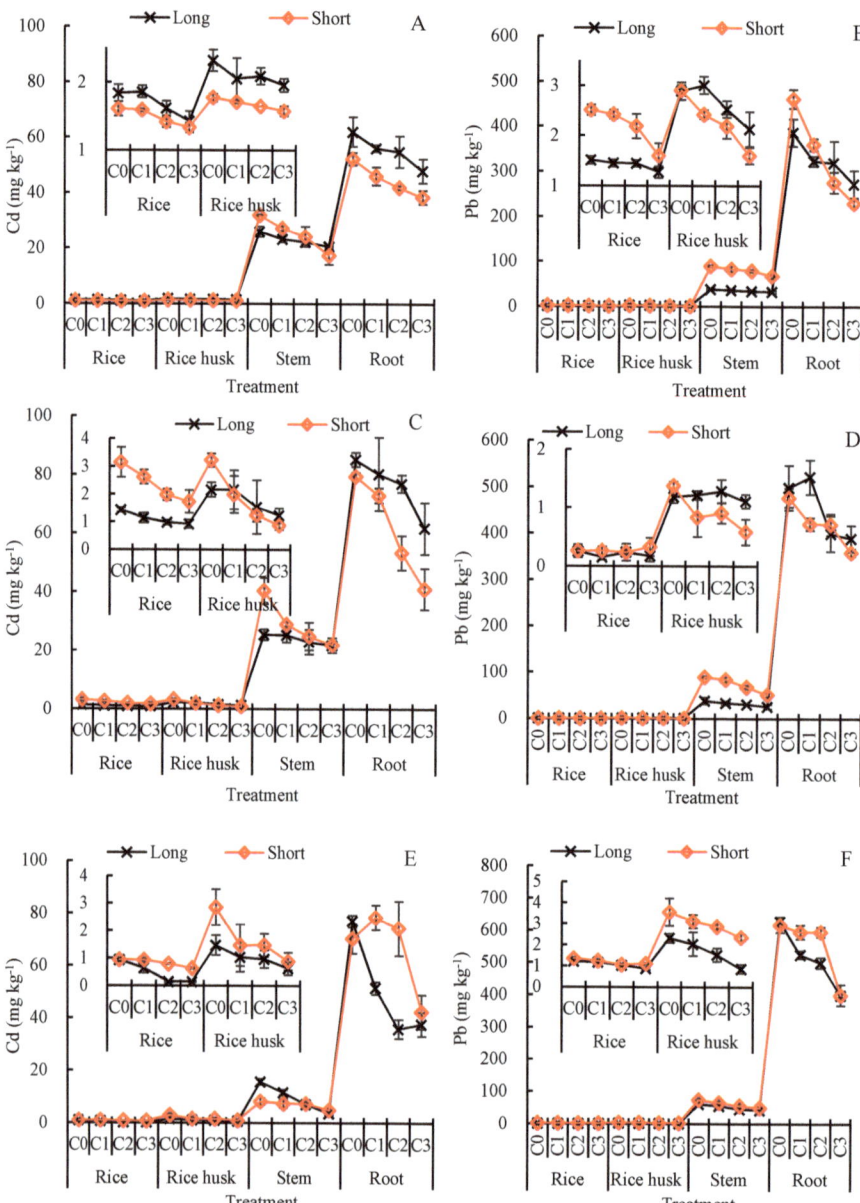

Figure 3. The effect of long- and short-term, increasing biochar application rates [0 (C0), 10 (C1), 20 (C2) and 40 (C3) t ha^{-1}] on Cd and Pb distribution in rice over time ((**A**): 2016 Cd; (**B**): 2016 Pb; (**C**): 2017 Cd; (**D**): 2017 Pb; (**E**): 2018 Cd; (**F**): 2018 Pb; Long: long-term; Short: short-term). Error bars represent standard deviation of the mean ($n = 3$).

3.4. Biochar Effects on Soil Enzyme Activity and Microbial Diversity

Soil enzyme activities are key indicators of ecological change, and are particularly sensitive to anthropogenic modifications of heavy metal contaminated soils. Enzymatic activity alterations, due to soil amendment applications, are directly expressed in strength of biochemical reactions and their associated implications within soils [50]. In particular, soil oxidoreductase enzymatic activities are sensitive to change, and include alkaline phosphatase, dehydrogenase, urease and sucrase enzymatic activities. Specifically, Tabatabai suggest that alkaline phosphatase describes a broad group of enzymes that catalyze the hydrolysis of both ester and anhydride organic P, leading to increased inorganic P availability [51]. Dehydrogenase activity is typically considered a measure of general microbial activity [52]. Urease activity has long been known to hydrolyze urea to ammonium [53], thus increasing N availability to plants. And, sucrase activity is involved in sucrose degradation as well as direct SOM metabolism, which would enhance nutrient availability [54].

Soil enzymatic activities, as a function of biochar application rate in the long- and short-term, are shown in Figure 4. Biochar applications led to increased enzymatic activities in this Cd and Pb contaminated soil. Increasing biochar application rate increased: (a) phosphatase enzyme activity by 8.8–104.9% (long-term) and 9.0–83.6% (short-term); (b) dehydrogenase enzyme activity by 9.3–118.5% (long-term) and 10.4–103.5% (short-term); and (c) urease and sucrase enzyme activity by 91.3% and 77.9% (long-term), 132.4% and 83.6% (short-term), respectively. One could construe increased enzymatic activities as indicators of positive ecosystem change.

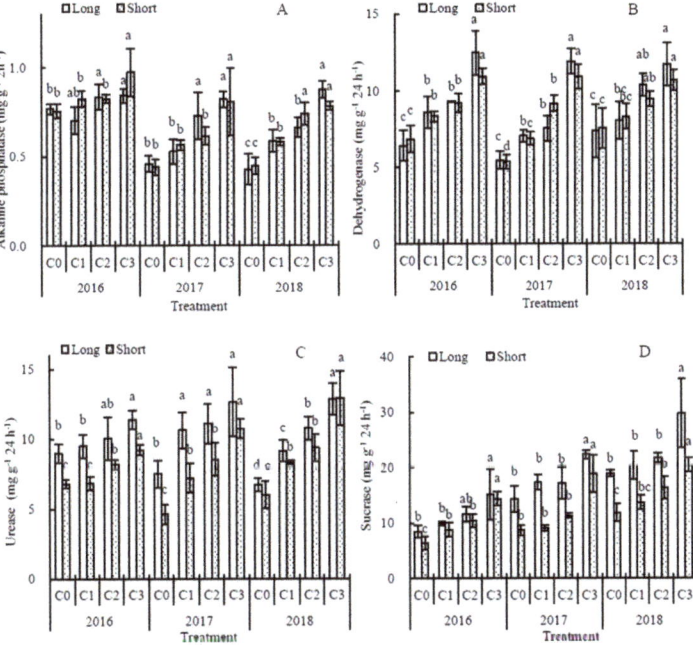

Figure 4. The effect of long- and short-term, increasing biochar application rates [0 (C0), 10 (C1), 20 (C2) and 40 (C3) t ha^{-1}] on (**A**) alkaline phosphatase, (**B**) dehydrogenase, (**C**) urease, and (**D**) sucrase activity. Different lower-case letters above error bars indicate significant differences between the biochar treatments in either the short- or long-term for a given year ($p < 0.05$, LSD post-hoc test). The error bars represent standard deviation of the mean ($n = 3$).

Others have also noted enzymatic activity changes associated with biochar application. Chen et al., showed that wood biochar (4%, *w:w*) increased sucrase activity by up to 12.5-fold as compared to a control, with a subsequent improvement in soil fertility status [55]. Biochar amendment caused both catalase and urease activities to gradually decrease up to 45 days, but then increased over longer timeframes [18]. However, others biochar studies have found opposite responses. Liu et al., reported decreased urease activity in biochar amended soils, likely the result of oxidative reactions with free radicals on biochar surfaces [56]. Huang et al., utilized rice straw biochar (5%, *w:w*), observing a decrease in urease and alkaline phosphatase activities [57]. Regardless, in our study, biochar applications appear to cause positive changes in soil enzymatic activity, and thus biochar use may lead to positive changes in ecosystem functionality.

Increasing biochar application rates tended to increase the Shannon microbial diversity index in both the long- and short-term (Figure 5A). The Shannon index increased by 8.2% and 3.4% (2016), and by 46.2% and 29.8% (2017) in the long- and short-term studies, with greatest changes associated with the highest biochar application rate. Similar Shannon diversity index changes associate with bacterial communities in biochar amended soils have been previously reported [58]. Zhang et al., found that biochar application (1.5%) had the highest richness estimators and Shannon diversity index in amended, Cd contaminated soil [59].

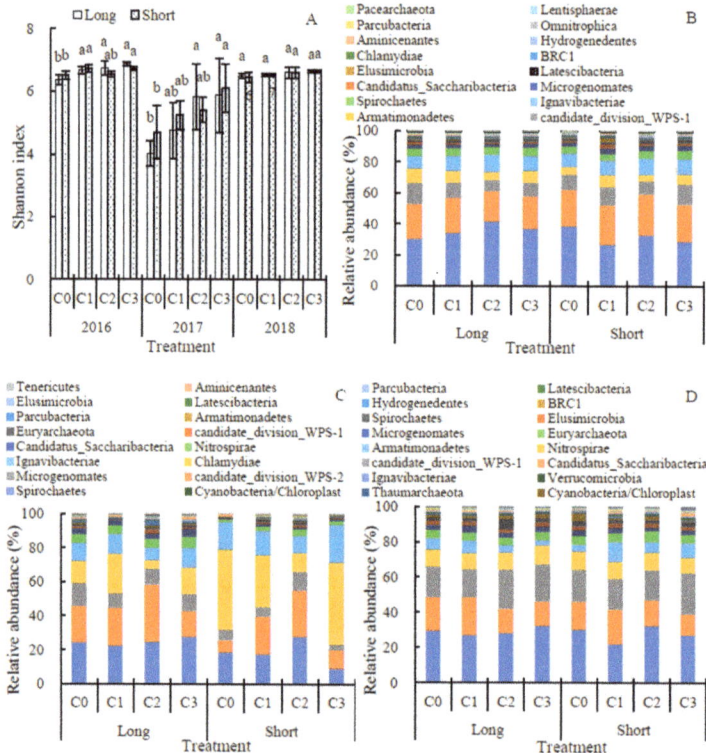

Figure 5. The effect of long- and short-term, increasing biochar application rates [0 (C0), 10 (C1), 20 (C2) and 40 (C3) t ha^{-1}] on changes of the (**A**) Shannon index and soil microbial diversity ((**B**): 2016; (**C**): 2017; (**D**): 2018). Different lower-case letters above error bars indicate significant differences between the biochar treatments in either the short- or long-term for a given year ($p < 0.05$, LSD post-hoc test). The error bars represent standard deviation of the mean ($n = 3$).

High-throughput PCR screening was used to further distinguish changes in microbial diversity (Figure 5B–D). During the studied years, microbial diversity increased by over 90% compared at phylum level. Steinbeiss et al., found similar responses due to biochar application. In the current study, biochar had the greatest effect on nine-phylum level microbial diversities in the short-term, significantly increasing them by 9.4–102.9% compared to the control (except *Chloroflexi* in the short-term in 2016) [60]. The nine-phylum level microbial diversities were slightly decreased by 6.4–46.9% in the long-term, except *Chloroflexi, Acidobacteria,* Unassigned and *Gemmatimonadetes*. Biochar (2%, $w:w$) has been shown to provide a positive effect on bacteria and invertebrate (such as earthworm) growth in heavy metal-contaminated agricultural soil in the short-term [2]. In a lab experiment, bacterial counts were increased by 149.4% compared with a control after 63 days following coconut shell biochar (5%, $w:w$) application to a Cd and Zn contaminated soil [38]. As with previous studies, in the current study, the soil microbial diversity was more greatly influenced in the short- versus long-term.

3.5. The Influence of Biochar on Soil and Plant Characteristics in Relation to Cd and Pb Contamination

Biochar application improved soil properties and reduced above-ground rice plant metals concentrations, with direct or indirect pathways illustrated via PCA and correlation analysis (Figure 6). The PCA results shown that biochar application rates were positively related to soil properties (e.g., SOM, pH), soil enzyme activity and soil moisture factors in PC1 (32.9%). Negative relationships between biochar applications and Cd and Pb concentrations in the soil exchangeable phase, and Cd and Pb concentrations in crops, were also observed in PC1. The factors of PC2 mainly included organic and residual Cd and Pb fractions and microbial diversity, which were positively related to biochar application rates. The factors of PC1 and PC2 explained 72% of the variability for all factors, indicating that biochar was one of the key factors affecting soil properties, various soil Cd and Pb fractions, and microbial diversity. Similarly, Tang et al., found that rice straw biochar application was effective at improving soil pH and SOM, leading to decreased Cd availability and subsequent increases in soil enzymatic activity [61]. Xing et al., showed that a sludge-based biochar increased the ratio of immobilized soil heavy metals, which positively correlated to the presence of *V. fischeri*, the growth of wheat, and the activities of other soil microbes [39].

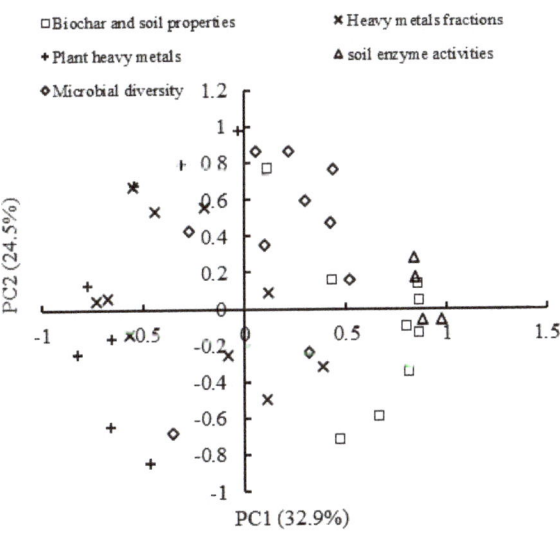

Figure 6. A principle component analysis.

4. Conclusions

Study results supported the hypothesis that wheat straw biochar could be used to decrease soil Cd and Pb bioavailability, decrease Cd and Pb uptake and translocation by rice, and improve a contaminated soil's overall bio-physicochemical characteristics. Biochar improved soil moisture content, increased soil pH, SOM, and reduced Cd and Pb bioavailability. Furthermore, biochar transformed bioavailable Cd and Pb to residual (i.e., unavailable) phases. In turn, biochar helped reduce the BCF and TF within rice, and reduced grain Cd and Pb concentrations to near or below China's national safety concentrations. Other positive responses were observed with respect to greater soil enzymatic activity and microbial community structure, although some positive changes were more evident over shorter (versus longer) time periods following biochar application. Future research should focus on atomic-level mechanisms (e.g., synchrotron-based studies) by which biochar sorbs and stabilizes long-term, in-situ heavy metals over longer time periods and in different soil types.

Supplementary Materials: The following are available online at http://www.mdpi.com/2076-3298/7/7/53/s1, Figure S1: the schematic diagram fo the acceptor cell title, Figure S2: The effect of long- and short-term, increasing biochar application rates [0(C0), 10 (C1), 20 (C2) and 40 (C3) t ha^{-1}] on water-soluble Cd and Pb with in-situ different time (A: Cd long-term; B: Cd short-term; C: Pb long-term; D: Pb short-term), Table S1: Basic paddy soil (0 to 15 cm depth) and biochar (g kg^{-1}) properties, Table S2: Bioconcentration (BCF) and translocation (TF) factors associated with biochar amend soil (biochar rates = 0(C0), 10 (C1), 20 (C2) and 40 (C3) t ha^{-1}).

Author Contributions: Conceptualization, L.C., L.L. and J.Y.; writing—original draft preparation, L.C. and R.B.; formal analysis, L.C., G.Q. and Y.L.; investigation, Y.L.; resources, L.C. and H.W.; review and editing, J.A.I. and L.C.; visualization, H.W.; supervision, L.L. and J.Y.; funding acquisition, L.C. and J.Y. All authors have read and agreed to the published version of the manuscript.

Funding: This study received funding from the National Natural Science Foundation of China: 41501339, National Natural Science Foundation of China: 21677119, Jiangsu Province Science Foundation for Youths: BK20140468 and from the QingLan Project.

Conflicts of Interest: The authors declare no conflict of interest.

References

1. Cui, L.; Noerpel, M.; Scheckel, K.G.; Ippolito, J.A. Wheat straw biochar reduces environmental cadmium bioavailability. *Environ. Int.* **2019**, *126*, 69–75. [CrossRef] [PubMed]
2. Hale, S.E.; Jensen, J.; Jakob, L.; Oleszczuk, P.; Hartnik, T.; Henriksen, T.; Okkenhaug, G.; Martinsen, V.; Cornelissen, G. Short-Term Effect of the Soil Amendments Activated Carbon, Biochar, and Ferric Oxyhydroxide on Bacteria and Invertebrates. *Environ. Sci. Technol.* **2013**, *47*, 8674–8683. [CrossRef] [PubMed]
3. MEP; MLR. The national soil contamination survey. *Natl. Soil Pollut. Surv. Bull.* **2014**, *4*, 1–5.
4. Shen, Z.; Jin, F.; O'Connor, D.; Hou, D. Solidification/Stabilization for Soil Remediation: An Old Technology with New Vitality. *Environ. Sci. Technol.* **2019**, *53*, 11615–11617. [CrossRef]
5. Wang, L.; Yu, K.; Li, J.; Tsang, D.C.W.; Poon, C.S.; Yoo, J.; Baek, K.; Ding, S.; Hou, D.; Dai, J. Low-carbon and low-alkalinity stabilization/solidification of high-Pb contaminated soil. *Chem. Eng. J.* **2018**, *351*, 418–427. [CrossRef]
6. Senneca, O.; Cortese, L.; Di Martino, R.; Fabbricino, M.; Ferraro, A.; Race, M.; Scopino, A. Mechanisms affecting the delayed efficiency of cement based stabilization/solidification processes. *J. Clean. Prod.* **2020**, *261*, 121230. [CrossRef]
7. Cui, L.Q.; Li, L.Q.; Zhang, A.F.; Pan, G.X.; Bao, D.D.; Chang, A. Biochar Amendment Greatly Reduces Rice Cd Uptake in a Contaminated Paddy Soil: A Two-Year Field Experiment. *Bioresources* **2011**, *6*, 2605–2618.
8. Ippolito, J.A.; Laird, D.A.; Busscher, W.J. Environmental Benefits of Biochar. *J. Environ. Qual.* **2012**, *41*, 967–972. [CrossRef]
9. Lehmann, J.; Rillig, M.C.; Thies, J.E.; Masiello, C.A.; Hockaday, W.C.; Crowley, D.E. Biochar effects on soil biota—A review. *Soil Biol. Biochem.* **2011**, *43*, 1812–1836. [CrossRef]
10. Li, X.; Wang, C.; Zhang, J.; Liu, J.; Liu, B.; Chen, G. Preparation and application of magnetic biochar in water treatment: A critical review. *Sci. Biochem. Environ.* **2020**, *711*, 134847. [CrossRef]

11. Khan, Z.H.; Gao, M.; Qiu, W.; Islam, M.S.; Song, Z. Mechanisms for cadmium adsorption by magnetic biochar composites in an aqueous solution. *Chemosphere* **2020**, *246*, 125701. [CrossRef] [PubMed]
12. Hemavathy, R.V.; Kumar, P.S.; Kanmani, K.; Jahnavi, N. Adsorptive separation of Cu(II) ions from aqueous medium using thermally/chemically treated Cassia fistula based biochar. *J. Clean. Prod.* **2020**, *249*, 119390. [CrossRef]
13. Zhao, L.; Nan, H.; Kan, Y.; Xu, X.; Qiu, H.; Cao, X. Infiltration behavior of heavy metals in runoff through soil amended with biochar as bulking agent. *Environ. Pollut.* **2019**, *254*, 113114. [CrossRef]
14. Qin, P.; Wang, H.; Yang, X.; He, L.; Müller, K.; Shaheen, S.M.; Xu, S.; Rinklebe, J.; Tsang, D.C.W.; Ok, Y.S.; et al. Bamboo- and pig-derived biochars reduce leaching losses of dibutyl phthalate, cadmium, and lead from co-contaminated soils. *Chemosphere* **2018**, *198*, 450–459. [CrossRef] [PubMed]
15. Ippolito, J.A.; Berry, C.M.; Strawn, D.G.; Novak, J.M.; Levine, J.; Harley, A. Biochars Reduce Mine Land Soil Bioavailable Metals. *J. Environ. Qual.* **2017**, *46*, 411–419. [CrossRef]
16. Chen, D.; Wang, X.; Wang, X.; Feng, K.; Su, J.; Dong, J. The mechanism of cadmium sorption by sulphur-modified wheat straw biochar and its application cadmium-contaminated soil. *Sci. Total Environ.* **2020**, *714*, 136550. [CrossRef]
17. Yin, D.; Wang, X.; Chen, C.; Peng, B.; Tan, C.; Li, H. Varying effect of biochar on Cd, Pb and As mobility in a multi-metal contaminated paddy soil. *Chemosphere* **2016**, *152*, 196–206. [CrossRef]
18. Tu, C.; Wei, J.; Guan, F.; Liu, Y.; Sun, Y.; Luo, Y. Biochar and bacteria inoculated biochar enhanced Cd and Cu immobilization and enzymatic activity in a polluted soil. *Environ. Int.* **2020**, *137*, 105576. [CrossRef]
19. Bian, R.; Joseph, S.; Cui, L.; Pan, G.; Li, L.; Liu, X.; Zhang, A.; Rutlidge, H.; Wong, S.; Chia, C. A three-year experiment confirms continuous immobilization of cadmium and lead in contaminated paddy field with biochar amendment. *J. Hazard. Mater.* **2014**, *272*, 121–128. [CrossRef]
20. Zhang, L.; Tang, S.; Jiang, C.; Jiang, X.; Guan, Y. Simultaneous and Efficient Capture of Inorganic Nitrogen and Heavy Metals by Polyporous Layered Double Hydroxide and Biochar Composite for Agricultural Nonpoint Pollution Control. *ACS Appl. Mater. Interfaces* **2018**, *10*, 43013–43030. [CrossRef]
21. Mujtaba Munir, M.A.; Liu, G.; Yousaf, B.; Ali, M.U.; Abbas, Q.; Ullah, H. Synergistic effects of biochar and processed fly ash on bioavailability, transformation and accumulation of heavy metals by maize (*Zea mays* L.) in coal-mining contaminated soil. *Chemosphere* **2020**, *240*, 124845. [CrossRef] [PubMed]
22. Cui, L.; Pan, G.; Li, L.; Yan, J.; Zhang, A.; Bian, R.; Chang, A. The Reduction of Wheat Cd Uptake in Contaminated Soil Via Biochar Amendment: A Two-Year Field Experiment. *Bioresources* **2012**, *7*, 5666–5676. [CrossRef]
23. Cui, L.; Pan, G.; Li, L.; Bian, R.; Liu, X.; Yan, J.; Quan, G.; Ding, C.; Chen, T.; Liu, Y.; et al. Continuous immobilization of cadmium and lead in biochar amended contaminated paddy soil: A five-year field experiment. *Ecol. Eng.* **2016**, *93*, 1–8. [CrossRef]
24. Gong, Z.; Lei, W.; Chen, Z.; Gao, Y.; Zeng, S.; Zhang, G.; Xiao, D.; Li, S. Chinese Soil Taxonomy. *Bull. Chin. Acad. Sci.* **2007**, *21*, 36–38.
25. Lu, R. Methods of inorganic pollutants analysis. In *Soil and Agro-Chemical Analysis Methods*; Agricultural Science and Technology Press: Beijing, China, 2000; pp. 205–266.
26. Parat, C.; Pinheiro, J.P. ISIDORE, a probe for in situ trace metal speciation based on Donnan membrane technique with related electrochemical detection part 1: Equilibrium measurements. *Anal. Chim. Acta* **2015**, *896*, 1–10. [CrossRef]
27. Ure, A.M.; Quevauviller, P.; Muntau, H.; Griepink, B. Speciation of Heavy Metals in Soils and Sediments. An Account of the Improvement and Harmonization of Extraction Techniques Undertaken Under the Auspices of the BCR of the Commission of the European Communities. *Int. J. Environ. Anal. Chem.* **1993**, *51*, 135–151. [CrossRef]
28. Wang, Y.-P.; Shi, J.-Y.; Qi, L.; Chen, X.-C.; Chen, Y.-X. Heavy metal availability and impact on activity of soil microorganisms along a Cu/Zn contamination gradient. *J. Environ. Sci.* **2007**, *19*, 848–853. [CrossRef]
29. Cui, L.; Yan, J.; Yang, Y.; Li, L.; Quan, G.; Ding, C.; Chen, T.; Fu, Q.; Chang, A. Influence of biochar on microbial activities of heavy metals contaminated paddy fields. *BioResources* **2013**, *8*, 5536–5548. [CrossRef]
30. Gong, X.; Jiang, Y.; Zheng, Y.; Chen, X.; Li, H.; Hu, F.; Liu, M.; Scheu, S. Earthworms differentially modify the microbiome of arable soils varying in residue management. *Soil Biol. Biochem.* **2018**, *121*, 120–129. [CrossRef]

31. van Leeuwen, H.P.; Town, R.M.; Buffle, J.; Cleven, R.F.M.J.; Davison, W.; Puy, J.; van Riemsdijk, W.H.; Sigg, L. Dynamic Speciation Analysis and Bioavailability of Metals in Aquatic Systems. *Environ. Sci. Technol.* **2005**, *39*, 8545–8556. [CrossRef]
32. Fan, Q.; Sun, J.; Quan, G.; Yan, J.; Gao, J.; Zou, X.; Cui, L. Insights into the effects of long-term biochar loading on water-soluble organic matter in soil: Implications for the vertical co-migration of heavy metals. *Environ. Int.* **2020**, *136*, 105439. [CrossRef]
33. Xu, Y.; Seshadri, B.; Sarkar, B.; Wang, H.; Rumpel, C.; Sparks, D.; Farrell, M.; Hall, T.; Yang, X.; Bolan, N. Biochar modulates heavy metal toxicity and improves microbial carbon use efficiency in soil. *Sci. Total Environ.* **2018**, *621*, 148–159. [CrossRef] [PubMed]
34. Wang, Y.; Zhong, B.; Shafi, M.; Ma, J.; Guo, J.; Wu, J.; Ye, Z.; Liu, D.; Jin, H. Effects of biochar on growth, and heavy metals accumulation of moso bamboo (Phyllostachy pubescens), soil physical properties, and heavy metals solubility in soil. *Chemosphere* **2019**, *219*, 510–516. [CrossRef] [PubMed]
35. Zhang, Y.; Chen, Z.; Xu, W.; Liao, Q.; Zhang, H.; Hao, S.; Chen, S. Pyrolysis of various phytoremediation residues for biochars: Chemical forms and environmental risk of Cd in biochar. *Bioresour. Technol.* **2020**, *299*, 122581. [CrossRef]
36. Chen, Q.; Dong, J.; Yi, Q.; Liu, X.; Zhang, J.; Zeng, Z. Proper Mode of Using Rice Straw Biochar To Treat Cd-Contaminated Irrigation Water in Mining Regions Based on a Multiyear in Situ Experiment. *ACS Sustain. Chem. Eng.* **2019**, *7*, 9928–9936. [CrossRef]
37. Wang, Y.; Xu, Y.; Li, D.; Tang, B.; Man, S.; Jia, Y.; Xu, H. Vermicompost and biochar as bio-conditioners to immobilize heavy metal and improve soil fertility on cadmium contaminated soil under acid rain stress. *Sci. Total Environ.* **2018**, *621*, 1057–1065. [CrossRef]
38. Liu, H.; Xu, F.; Xie, Y.; Wang, C.; Zhang, A.; Li, L.; Xu, H. Effect of modified coconut shell biochar on availability of heavy metals and biochemical characteristics of soil in multiple heavy metals contaminated soil. *Sci. Total Environ.* **2018**, *645*, 702–709. [CrossRef]
39. Xing, J.; Li, L.; Li, G.; Xu, G. Feasibility of sludge-based biochar for soil remediation: Characteristics and safety performance of heavy metals influenced by pyrolysis temperatures. *Ecotoxicol. Environ. Saf.* **2019**, *180*, 457–465. [CrossRef]
40. Beesley, L.; Inneh, O.S.; Norton, G.J.; Moreno-Jimenez, E.; Pardo, T.; Clemente, R.; Dawson, J.J.C. Assessing the influence of compost and biochar amendments on the mobility and toxicity of metals and arsenic in a naturally contaminated mine soil. *Environ. Pollut.* **2014**, *186*, 195–202. [CrossRef]
41. Novak, J.M.; Busscher, W.J.; Watts, D.W.; Amonette, J.E.; Ippolito, J.A.; Lima, I.M.; Gaskin, J.W.; Das, K.C.; Steiner, C.; Ahmedna, M. Biochars impact on soil moisture storage in an Ultisol and two Aridisols. *Soil Sci.* **2012**, *177*, 310–320. [CrossRef]
42. Ippolito, J.A.; Stromberger, M.E.; Lentz, R.D.; Dungan, R.S. Hardwood biochar influences calcareous soil physicochemical and microbiological status. *J. Environ. Qual.* **2014**, *43*, 681–689. [CrossRef] [PubMed]
43. Ippolito, J.A.; Stromberger, M.E.; Lentz, R.D.; Dungan, R.S. Hardwood biochar and manure co-application to a calcareous soil. *Chemosphere* **2016**, *142*, 84–91. [CrossRef] [PubMed]
44. Lentz, R.D.; Ippolito, J.A.; Lehrsch, G.A. Biochar, Manure, and Sawdust Alter Long-Term Water Retention Dynamics in Degraded Soil. *Soil Sci. Soc. Am. J.* **2019**, *83*, 1491–1501. [CrossRef]
45. Liu, S.; Pu, S.; Deng, D.; Huang, H.; Yan, C.; Ma, H.; Razavi, B.S. Comparable effects of manure and its biochar on reducing soil Cr bioavailability and narrowing the rhizosphere extent of enzyme activities. *Environ. Int.* **2020**, *134*, 105277. [CrossRef] [PubMed]
46. Wang, F.; Wang, H.; Al-Tabbaa, A. Leachability and heavy metal speciation of 17-year old stabilised/solidified contaminated site soils. *J. Hazard. Mater.* **2014**, *278*, 144–151. [CrossRef]
47. NHFPC; CFDA. *National Standards for Food Safety Limits for Contaminants in Food*; China Food and Drug Administration Press: Beijing, China, 2017; pp. 2–5.
48. Turan, V.; Khan, S.A.; Mahmood ur, R.; Iqbal, M.; Ramzani, P.M.A.; Fatima, M. Promoting the productivity and quality of brinjal aligned with heavy metals immobilization in a wastewater irrigated heavy metal polluted soil with biochar and chitosan. *Ecotoxicol. Environ. Saf.* **2018**, *161*, 409–419. [CrossRef]
49. Puga, A.P.; Abreu, C.A.; Melo, L.C.A.; Beesley, L. Biochar application to a contaminated soil reduces the availability and plant uptake of zinc, lead and cadmium. *J. Environ. Manag.* **2015**, *159*, 86–93. [CrossRef]

50. Bandara, T.; Franks, A.E.; Xu, J.; Bolan, N.; Wang, H.; Tang, C. Chemical and biological immobilization mechanisms of potentially toxic elements in biochar-amended soils. *Crit. Rev. Environ. Sci. Technol.* **2019**, *50*, 903–978. [CrossRef]
51. Tabatabai, M.A. Soil enzymes. In *Methods of Soil Analysis, Part 2. Microbiological and Biochemical Properties*; SSSA Book Series No. 5; Weaver, R.W., Angle, J.S., Bottomley, P.S., Eds.; Soil Science Society of America: Madison, WI, USA, 1994; pp. 775–833.
52. Moeskops, B.; Sukristiyonubowo; Buchan, D.; Sleutel, S.; Herawaty, L.; Husen, E.; Saraswati, R.; Setyorini, D.; De Neve, S. Soil microbial communities and activities under intensive organic and conventional vegetable farming in West Java, Indonesia. *Appl. Soil Ecol.* **2010**, *45*, 112–120. [CrossRef]
53. Lloyd, A.B.; Sheaffe, M.J. Urease activity in soils. *Plant Soil* **1973**, *39*, 71–80. [CrossRef]
54. Liu, J.G.; Zhang, W.; Yan-Bin, L.I.; Sun, Y.Y.; Bian, X.M. Effects of Long-Term Continuous Cropping System of Cotton on Soil Physical-Chemical Properties and Activities of Soil Enzyme in Oasis in Xinjiang. *Sci. Agric. Sin.* **2009**, *42*, 725–733.
55. Chen, H.; Yang, X.; Wang, H.; Sarkar, B.; Shaheen, S.M.; Gielen, G.; Bolan, N.; Guo, J.; Che, L.; Sun, H.; et al. Animal carcass- and wood-derived biochars improved nutrient bioavailability, enzyme activity, and plant growth in metal-phthalic acid ester co-contaminated soils: A trial for reclamation and improvement of degraded soils. *J. Environ. Manag.* **2020**, *261*, 110246. [CrossRef] [PubMed]
56. Liu, Y.; Dai, Q.; Jin, X.; Dong, X.; Peng, J.; Wu, M.; Liang, N.; Pan, B.; Xing, B. Negative Impacts of Biochars on Urease Activity: High pH, Heavy Metals, Polycyclic Aromatic Hydrocarbons, or Free Radicals? *Environ. Sci. Technol.* **2018**, *52*, 12740–12747. [CrossRef]
57. Huang, D.; Liu, L.; Zeng, G.; Xu, P.; Huang, C.; Deng, L.; Wang, R.; Wan, J. The effects of rice straw biochar on indigenous microbial community and enzymes activity in heavy metal-contaminated sediment. *Chemosphere* **2017**, *174*, 545–553. [CrossRef] [PubMed]
58. Wong, J.T.F.; Chen, X.; Deng, W.; Chai, Y.; Ng, C.W.W.; Wong, M.H. Effects of biochar on bacterial communities in a newly established landfill cover topsoil. *J. Environ. Manag.* **2019**, *236*, 667–673. [CrossRef] [PubMed]
59. Zhang, M.; Wang, J.; Bai, S.H.; Zhang, Y.; Teng, Y.; Xu, Z. Assisted phytoremediation of a co-contaminated soil with biochar amendment: Contaminant removals and bacterial community properties. *Geoderma* **2019**, *348*, 115–123. [CrossRef]
60. Steinbeiss, S.; Gleixner, G.; Antonietti, M. Effect of biochar amendment on soil carbon balance and soil microbial activity. *Soil Biol. Biochem.* **2009**, *41*, 1301–1310. [CrossRef]
61. Tang, J.; Zhang, L.; Zhang, J.; Ren, L.; Zhou, Y.; Zheng, Y.; Luo, L.; Yang, Y.; Huang, H.; Chen, A. Physicochemical features, metal availability and enzyme activity in heavy metal-polluted soil remediated by biochar and compost. *Sci. Total Environ.* **2020**, *701*, 134751. [CrossRef]

© 2020 by the authors. Licensee MDPI, Basel, Switzerland. This article is an open access article distributed under the terms and conditions of the Creative Commons Attribution (CC BY) license (http://creativecommons.org/licenses/by/4.0/).

Article

Influence of Biochar Derived Nitrogen on Cadmium Removal by Ryegrass in a Contaminated Soil

João Antonangelo * and Hailin Zhang

Plant and Soil Sciences Department, Oklahoma State University, Stillwater, OK 74078, USA; hailin.zhang@okstate.edu
* Correspondence: joao.antonangelo@okstate.edu; Tel.: +1-405-780-3950

Abstract: Little is known about the effect of nitrogen (N) application via biochar on the removal of trace elements by crops, and the effects with chemical fertilizers are inconsistent. We determined, from a previous study, the influence of increased N addition via biochars produced from switchgrass (SGB) and poultry litter (PLB) on cadmium (Cd) removal by ryegrass. The biochar rates of 0, 0.5, 1, 2, and 4% w/w were applied to a Cd-contaminated soil before seeding in a potting experiment with a complete randomized block design (CRBD). Ryegrass yield and N and Cd removed by harvest were strongly related ($p < 0.05$). The ryegrass yields increased up to 1% of PLB, and Cd removal was also the highest at 1% of PLB. The biomass of ryegrass roots increased with Cd accumulation ($p < 0.05$). Overall, the Cd transfer factor (TF) from ryegrass roots to shoots increased when up to 206 ± 38 kg N ha^{-1} was removed in ryegrass shoots ($p < 0.0001$). The application of PLB up to 1% might be a viable option since it is a practical rate for handling operations requiring less volume of material than SGB. Additionally, the Cd concentration in the aboveground forage remained acceptable for grazing cattle. Future studies are encouraged to evaluate different sources of N fertilizers affecting Cd uptake on cash crops.

Keywords: cadmium; biochar application rates; metal accumulation; nitrogen; ryegrass

Citation: Antonangelo, J.; Zhang, H. Influence of Biochar Derived Nitrogen on Cadmium Removal by Ryegrass in a Contaminated Soil. *Environments* **2021**, *8*, 11. https://doi.org/10.3390/environments8020011

Academic Editor: Dionisios Gasparatos
Received: 7 January 2021
Accepted: 2 February 2021
Published: 8 February 2021

Publisher's Note: MDPI stays neutral with regard to jurisdictional claims in published maps and institutional affiliations.

Copyright: © 2021 by the authors. Licensee MDPI, Basel, Switzerland. This article is an open access article distributed under the terms and conditions of the Creative Commons Attribution (CC BY) license (https://creativecommons.org/licenses/by/4.0/).

1. Introduction

Human activities have gradually transferred many toxic metals from the earth's crust to the environment, resulting in the spread and contamination of toxic metals in the ecosystem [1–4]. The metals originate from many sources including mining activities, industrial waste disposals, paints, and gasoline additives that lead to physical and chemical processes such as leaching and oxidation thus causing the accumulation of metals in the soil. Toxic metal pollution has become a serious problem worldwide in the last decade. Cadmium (Cd) is considered the most toxic element among toxic metals for living organisms and has been detected in some agricultural lands [1]. Cadmium accumulates in plants and animals and threatens their health when entering through the food chain [5,6]. Thus, it is important but challenging to remediate Cd-contaminated soils worldwide.

The Tar Creek area is designated as a superfund site located in the tri-state regions of Oklahoma, Kansas, and Missouri by US Environmental Protection Agency (EPA), and is one of the most polluted sites in the world [7]. A Superfund site is an abandoned hazardous waste site in the United States subject to the Comprehensive Environmental Response, Compensation and Liability Act (CERCLA), which allows federal funding to be spent on investigating and remediating the environmental contamination at the designated site. The Tar Creek area was mined for approximately 70 years starting from the beginning of the 1900s. Large quantities of coarse material contaminated with Cd were left on the ground surface in piles due to the mining and milling operations. Such a coarse material is gravel-like waste well known as "chat pile." The large size and the considerable amount of chat piles resulted in Cd contamination in the surrounding areas due to wind blowing and the chat materials used as gravel. Reclamations of contaminated lands have been attempted by the US Government and the Quapaw Tribe. Although the status of Cd contamination in the Tar Creek has been addressed in numerous studies [8–10], with some of them elucidating the use of biochars for Cd remediation and phytoavailability reduction [11], there is still a lack of research evaluating the effect of nitrogen (N) application via biochars on the accumulation of Cd by crops.

Nitrogen is the most important essential nutrient for plants, and N application is the biggest contributor to biomass production [12]. Although some studies suggested that Cd concentration in plants decreased as their biomass increased by N application [13,14], other works concluded that N fertilizer not only improved plants' biomass but also increased their Cd concentration [12,15–17]. In an experiment conducted by Symanowicz et al. [18], N fertilization significantly contributed to increase Cd accumulation in the aboveground portion (shoots) of eastern galega. Thus, the effect of N application on Cd accumulation remains controversial [12]. In general, N fertilization often decreases soil pH, which increases the solubility and mobility of Cd in the soil [19,20]. Ji et al. [12] also highlighted a more direct effect of N fertilization in Cd uptake by suggesting a mechanism in which the active uptake and influx rate of Cd^{2+} into the roots of Italian ryegrass (*Lolium multifolorum* Lam.) were enhanced by the greater affinity of the membrane transporter to Cd^{2+} as a consequence of the urea application.

Although the studies mentioned above have evaluated the effect of N on Cd bioavailability and plant uptake, all of them were performed through the addition of chemical fertilizers. Therefore, the evaluation of N addition using biochar aiming the same goal makes our research unique since there might be a compound effect of biochar properties, such as alkalinity and surface functional groups, also affecting Cd uptake, not just the N addition itself.

Perennial ryegrasses (*Lolium perenne*) are widely used in grazed pastures due to its potential to reduce nitrate (NO_3^-) leaching [21], which makes such grass a potential scavenger of soil N. Ryegrass is also known for its efficient phytoaccumulation of toxic metals such as Cd [11]. These characteristics make ryegrass a promising tool to study the N and Cd accumulation relationship in plant tissues. It is a gramineous grass with fast growth, high yield potential, strong resistance to toxic metals, and able to normally grow in tailing areas where toxic metal pollution is severe and the environment is harsh [12]. Moreover, Mongkhonsin et al. [22] pointed out that ryegrass shoots had the highest Cd enrichment of up to fourfold of what is commonly found in other plant species that are hyperaccumulators (*Solanum nigrum* and *Indian mustard*); and Italian ryegrass had the strongest tolerance and accumulation capacity of Cd among eight C3 herbage grass species. Therefore, the perennial ryegrass with high biomass yields is also an appropriate species to use for the remediation of Cd-contaminated soils [23].

Our objective was to determine the influence of increased N addition via biochar derived from two feedstocks on Cd removal by ryegrass. This study is derived from the work of Antonangelo and Zhang [11] and data were reanalyzed to evaluate the N and Cd removal instead of their simple uptake. Since current literature is undoubtedly deficient in such information, we believe that our research has the potential to expand from small-scale to field-scale studies. This will serve as a multi-purpose biochar application to increase forage biomass and Cd uptake for phytoremediation purposes, at the same time immobilizing a certain amount of bioavailable Cd in the soil, as observed with the previous study of Antonangelo and Zhang [11], thus avoiding its leaching to the groundwater.

This would be a pioneering work for other studies focusing on N rates from biochars and/or the combination of biochar + N fertilizers to increase toxic metal phytoaccumulation while immobilizing the metal in the soil-plant system. This multi-purpose aspect contributes to the sustainable use of resources, since only immobilization is not enough if there is a need for disposal, land application, or landfill. Hence, if animal grazing is needed in such contaminated areas, the risk of intoxication could be still avoided depending on the metal accumulation in shoots. A sustainable attitude (multi-purpose) like this might be useful for contaminated lands, such as the Tar Creek superfund site.

2. Materials and Methods

2.1. Biochar Production and Characteristics

The biochars used were converted from switchgrass (SG; *Panicum virgatum*) and poultry litter (PL) via slow pyrolysis at 700 °C. The feedstocks obtainment and the detailed description of their conversion into biochar products can be found in Antonangelo and Zhang [11]. Switchgrass- and poultry litter–derived biochars are referred to as SGB and PLB, respectively. The biochars coarse materials were ground with a mortar and pastel gently before sieved through 1 mm for further physicochemical analyses [24] and through 0.25 mm for the potting experiment. A full characterization of SGB and PLB including all physicochemical properties such as moisture, ash content, particle-size distribution, elemental composition, surface functional groups, chemical attributes, specific surface area, cation exchange capacity (CEC), and morphology can be found in [24]. For this study, it is useful to emphasize the ash, total carbon (TC), total nitrogen (TN), phosphorus (P), potassium (K), calcium (Ca), magnesium (Mg), and sulfur (S) contents, as presented in Table 1.

Table 1. Physicochemical properties of biochar derived from switchgrass (SGB) and poultry litter.

Property	Unit	SGB	PLB
Ash	%	4.4	45.9
TC	g kg^{-1}	314 ± 13	278 ± 14
TN	g kg^{-1}	7 ± 0	16 ± 1
P	g kg^{-1}	2 ± 0.2	40 ± 0.8
K	g kg^{-1}	4 ± 0.4	80 ± 0.6
Ca	g kg^{-1}	8 ± 1	50 ± 2
Mg	g kg^{-1}	3 ± 0	20 ± 0.1
S	g kg^{-1}	0.4 ± 0.1	10 ± 0.6

Numbers followed by "±" are the standard deviations of triplicates (n = 3).

2.2. Soil Sampling and Analysis

Cadmium contaminated soil samples were collected with a shovel (0 to 15 cm) from a residential yard near chat piles located in Picher, Ottawa County, Oklahoma. Soil sample preparation for analyses and for the potting experiment was described in Antonangelo and Zhang [11]. Before and after the pot experiment, soil samples were analyzed for DTPA–extractable Cd. Before the potting experiment, the total Cd content was determined by an inductively coupled plasma-atomic emission spectroscopy, ICP-AES (SPECTRO Analytical Instruments GmbH, Boschstr. 10, 47533 Kleve, Germany) after digestion by concentrated HNO_3 and H_2O_2 using EPA method 3050B [25]. The TN, and total and DTPA-extractable Cd before the potting experiment were respectively 1.7 ± 0.1 g kg^{-1}, 9 ± 0.6 mg kg^{-1}, and 1.84 ± 0.03 mg kg^{-1}. The total Cd content was about tenfold the maximum found in normal Oklahoma soils [26]. The DTPA-extractable Cd has been considered the most efficient method to predict the Cd phytoavailability to grasses, such as rice (*Oryza sativa*) [27] and perennial ryegrass [11]. Still, the threshold of DTPA-extractable Cd found by Wu et al. [27] was 0.03 to 0.16 mg kg^{-1} depending on the soil organic matter (SOM) content, which is more than 10 times lower than the content found in the Tar Creek soil of this study.

2.3. Potting Experiment

Plastic pots were filled with 1.2 kg of 2 mm sieved soils and amended with 0.0 (control), 0.5, 1.0, 2.0, and 4.0% (w/w) of 0.25 mm sieved SGB and PLB. The biochar-amended soils were incubated for approximately 30 days at 75% of field capacity before ryegrass sowing in each pot at 30 kg ha^{-1}. Ryegrass was grown for ~75 days in the biochar amended soils in an environmentally controlled growth chamber. The pots were arranged in a complete randomized block design (CRBD) with 3 replicates (n = 3) and rotated weekly to eliminate spatial variability in the chamber. All procedures and analyses conducted during the potting experiment are detailed in Antonangelo and Zhang [11].

2.4. Nitrogen and Cadmium Analysis in Ryegrass Shoots and Roots

After harvesting, ryegrass shoots and roots were separated, washed with D.I. water, and oven-dried to constant weight at 105 °C, and their weight was recorded. Dried plant materials were ground using a mechanical grinder and further analyses were followed. Ground plant materials were analyzed for Cd using nitric acid digestion and TN using dry combustion and determination via LECO, as previously described for soil. For Cd in plant tissues, 0.5 g of ground plant materials were predigested for 1h with 10 mL of trace metal grade HNO_3 in the HotBlock™ Environmental Express block digester (Environmental Express, 2345A Charleston Regional Parkway, Charleston, SC, USA), and the digestion products were then heated to 115 °C for 2 h and diluted with D.I. water to 50 mL [28]. Finally, the digested samples were filtered and analyzed for Cd by an ICP-AES.

The N and Cd concentration determined in plant tissues (uptake) was multiplied by their respective biomass to eliminate any dilution effect caused by over yield and the N and Cd in ryegrass shoots and/or roots were then treated as "removed" (shoots, when harvested) and/or "accumulated": $N/Cd\ removed/accumulated = (N/Cd\ concentration \times Yield)$.

The Cd transfer factor (TF) was calculated by considering the Cd removal and accumulation following: $TF = (Cd\ removed\ in\ shoots \div Cd\ accumulated\ in\ roots)$.

2.5. Statistical Analysis

Ryegrass N and Cd accumulation in ryegrass tissues and Cd TF from ryegrass shoots to roots, as described above, were subjected to a two-way ANOVA following the CRBD and the average of the treatments (biochar rates and biochar from different feedstocks) compared by the Tukey test at 5% ($p \leq 0.05$). Analysis of covariance (ANCOVA) was performed for the linear regressions between ryegrass yield and N accumulation and between N and Cd accumulations to plot the linear models with both biochars combined or for each biochar separately, depending on the results. Data analysis was performed using SAS version 9.4.

The maximum N removal from which the highest Cd TF is reached was determined by fitting the segmented linear-plateau response model using the NLIN (nonlinear) procedure of SAS version 9.4. Pearson's simple correlation for specific variables has also been carried out. Regression analyses and Pearson correlation were performed with the whole dataset of measurements (all replicate data) at α = 0.05 for both model fitting and equation coefficients. Graphs were plotted with the assistance of Excel.

3. Results

3.1. Ryegrass Production, and Cadmium × Nitrogen Accumulation Relationship

Ryegrass N and Cd accumulation are shown in Table 2. A two-way ANOVA was conducted to test the differences among biochar rates and between the two different feedstock-derived biochars for a given rate. This was because the interaction *rate (R) × biochar (B)* was significant ($p < 0.05$) in all cases: shoots, roots, and shoots + roots (Table 2). As highlighted in the previous work of Antonangelo and Zhang [11], the ryegrass shoots, roots, and whole plant (shoots + roots) yielded better when treated with PLB presenting, on average, an increase of 64, 51, and 59% respectively when compared to those treated with SGB (Table 2). Similarly, N and Cd accumulation had an overall increase of 88, 60, and 84%, and 71, 30, and 40% respectively, when PLB was applied (Table 2). At 0.5 and 1% of PLB application, the ryegrass shoots presented the highest N and Cd accumulation (Table 2), accompanied by their highest yields. This is the first indication that Cd accumulation increases as the ryegrass yield increase with N accumulation. The approximately 35 g Cd removed ha^{-1} in ryegrass shoots after PLB application at 0.5 and 1% (Table 2) is equivalent to a concentration of 3.4 ± 0.1 mg kg^{-1} (dry matter, DM) in our study, which is still far below the maximum tolerable concentration of 10 mg kg^{-1} for animal grazing [29]. The Cd accumulation in ryegrass plant parts decreased with biochar application rates regardless the feedstock from which the biochar was produced (Table 2). However, it must be pointed out that such reduction is not only a consequence of the reduced N removal and ryegrass yields at 4%

of biochar amendment but also due to a direct effect of the biochar properties with the potential to immobilize Cd in the soil, amongst them surface functional groups, alkalinity, organic carbon (OC), and CEC [24].

Table 2. Nitrogen (N) and cadmium (Cd) accumulation in ryegrass shoots, roots, and shoots + roots as a function of biochar (B) application rates (R).

Rate (%)	Shoots											
	N (kg ha^{-1})						Cd (g ha^{-1})					
	SGB		PLB		Mean		SGB		PLB		Mean	
0	221	aA	221	cA	221	a	30	aA	30	aA	30	a
0.5	113	bB	287	aA	200	b	15.2	bB	35.5	aA	25.4	a
1	99	bB	268	bA	183	b	14.3	bB	34.7	aA	24.5	a
2	72	cB	180	dA	126	c	10.6	bB	21.6	bA	16.1	b
4	19	dB	28	eA	24	d	2.3	cB	2.4	cA	2.3	c
Mean	105	B	197	A			14.5	B	24.8	A		
R			$p < 0.05$						$p < 0.05$			
B			$p < 0.05$						$p < 0.05$			
R × B			$p < 0.05$						$p < 0.05$			

Rate (%)	Roots											
	N (kg ha^{-1})						Cd (g ha^{-1})					
	SGB		PLB		Mean		SGB		PLB		Mean	
0	4.95	aA	4.95	cA	4.95	b	45.9	aA	45.9	cA	45.9	ab
0.5	3.84	bB	8.29	aA	6.06	a	60.5	aB	90.6	abA	75.5	a
1	3.42	bcB	7.11	bA	5.27	ab	61.7	aB	109.7	aA	85.7	a
2	2.52	cdB	4.91	cA	3.71	c	40.9	aA	52.7	bcA	46.8	ab
4	2.04	dA	1.49	dA	1.76	d	36.7	aA	15.3	cA	26	b
Mean	3.35	B	5.35	A			49.1	B	62.8	A		
R			$p < 0.05$						$p < 0.05$			
B			$p < 0.05$						$p < 0.05$			
R × B			$p < 0.05$						$p < 0.05$			

Rate (%)	Shoots + Roots											
	N (kg ha^{-1})						Cd (g ha^{-1})					
	SGB		PLB		Mean		SGB		PLB		Mean	
0	334	aA	334	bA	334	a	193.6	aA	193.6	bA	193.6	bc
0.5	180	bB	443	aA	312	ab	205.8	aB	343.7	aA	274.8	ab
1	157	bB	419	aA	288	b	208.4	aB	383.5	aA	296	a
2	114	cB	281	cA	198	c	142.1	abB	195.2	bA	168.6	c
4	43	dA	54	dA	48	d	75.6	bA	39.4	cA	57.5	d
Mean	166	B	306	A			165.1	B	231.1	A		
R			$p < 0.05$						$p < 0.05$			
B			$p < 0.05$						$p < 0.05$			
R × B			$p < 0.05$						$p < 0.05$			

SGB: switchgrass-derived biochar. PLB: poultry litter-derived biochar. Different lowercase letters in columns (within biochar rates) and uppercase letters in rows (same biochar rate) are significantly different at $p < 0.05$ (Tukey).

Table 3 shows the ANCOVA of linear regressions between ryegrass yield and N accumulation and between N and Cd accumulations plotted in Figure 1. Conversely to ANOVA, the interactions of *biochar × yield* and *biochar × nitrogen* from ANCOVA were not significant in ryegrass shoots and roots ($p > 0.05$), except for the whole plant when evaluating the *biochar × nitrogen* interaction in the relationship between N and Cd accumulations ($p < 0.05$) (Table 3). Therefore, most relationships were plotted with both biochars combined (Figure 1a–e).

Table 3. Analyses of covariance (ANCOVA) from linear regression models between nitrogen (N) and ryegrass yield and between cadmium (Cd) and N accumulations.

Factor	Shoots	Roots	Shoots + Roots
	N accumulated (y-axis)		
Biochar	$p = 0.9503$	$p = 0.4765$	$p = 0.5697$
Yield	$p < 0.0001$	$p = 0.0005$	$p < 0.0001$
Biochar × Yield	$p = 0.1767$	$p = 0.2452$	$p = 0.0731$
	Cd accumulated (y-axis)		
Biochar	$p = 0.1051$	$p = 0.4597$	$p = 0.4012$
Nitrogen	$p < 0.0001$	$p = 0.0003$	$p < 0.0001$
Biochar × Nitrogen	$p = 0.8557$	$p = 0.1298$	$p = 0.0104$

When $p \leq 0.05$ the effect test is significant for the designated factor. When the interaction *biochar × yield* or *nitrogen* is $p \leq 0.05$, results must be presented as two regression lines, one for each biochar.

Figure 1 clearly shows that N accumulation increases as a function of biochar application rates regardless of its feedstock, and such an accumulation contributes to ryegrass yield increases in shoots and roots. Concomitantly, the Cd accumulation of those plant parts follows their yields increment as well (Figure 1). This agrees with the findings of [12,15–18] who observed greater yields with increased N rates and consequently greater Cd uptake in the studied plants. In fact, N is the most limiting nutrient as far as yield increase is concerned and, in the case of our study, the same evaluations carried out for P and K showed a positive correlation with ryegrass yields but no relationship was found between P and K with Cd accumulation in any plant part (data not shown). Opposite to the increase of Cd accumulation in ryegrass plant parts, the Cd concentration in ryegrass shoots and roots, when their yields are not accounted for, decreased considerably with biochar application due to immobilization effect of phytoavailable Cd as a direct consequence of biochar properties, as observed in the previous study [11].

Curiously, the relationship between the yield of ryegrass roots and Cd accumulation was highly positive regardless of the biochar (Figure 2). There was an increment of 23 mg Cd accumulated ha^{-1} for every kg of ryegrass roots increase (Figure 2). The maximum accumulation in the range of 80 to 160 g Cd ha^{-1} is related to the PLB at 0.5 and 1% application rates (Table 2, and Figures 1 and 2). Feng et al. [30] also reported an increase in root yields (DM) of two ryegrass varieties when the levels of phytoavailable Cd were higher. This might be attributed to the fact that root density is higher in the presence of this toxic metal. However, Feng et al. [30] also observed that ryegrass roots although presenting higher enrichment ability for Cd, its transportability to shoots was poor. Not surprisingly, the accumulation of Cd in ryegrass roots was much higher than in shoots in the case of our study (Table 2 and Figure 1e). This larger accumulation of Cd in the roots is also a cause of the lower Cd transference to the shoots [31,32], as will be discussed further.

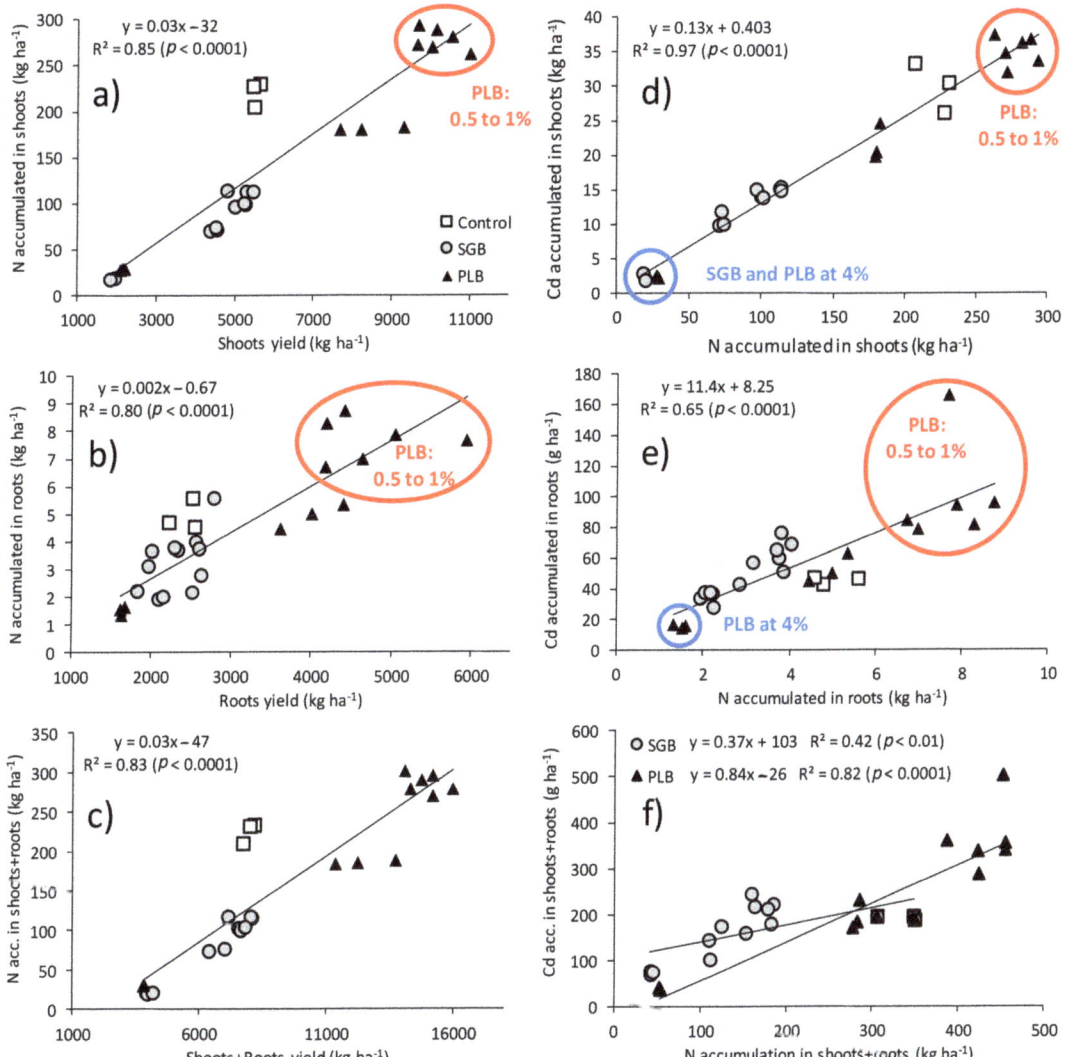

Figure 1. Relationships between ryegrass yield and N accumulation (**a–c**) and between nitrogen (N) and cadmium (Cd) accumulations in ryegrass shoots and roots (**d–f**) as functions of biochar application rates; acc.—accumulated.

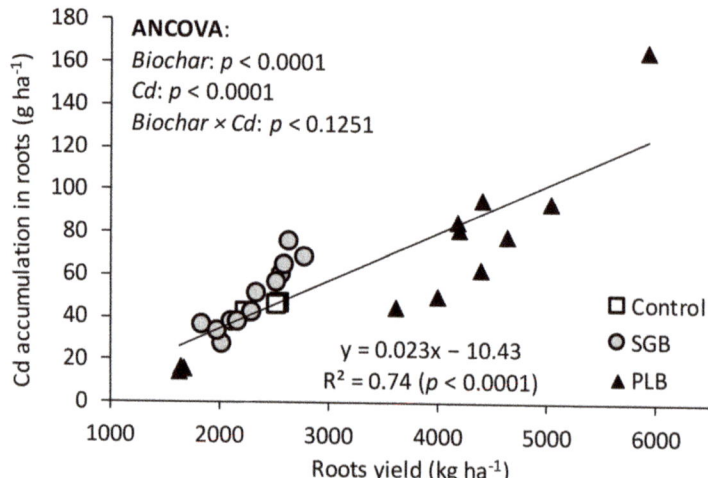

Figure 2. Relationship between root yields and cadmium (Cd) accumulation in ryegrass roots as a function of biochar application rates. ANCOVA: analysis of covariance. When the interaction biochar × Cd is $p \geq 0.05$, results of both biochars must be combined in a single linear regression.

3.2. Cadmium Bioavailability

Overall, DTPA-extractable Cd concentrations were higher for the control and decreased as the rates of biochars increased [11]. Previous studies also suggested that biochars incorporation to soils effectively decreased the Cd availability as assessed by DTPA [33–38]. In this study, we correlated the DTPA-extractable Cd from the biochar-treated soils with Cd and accumulation in ryegrass plant parts (Table 4). Some studies revealed that the dilution effect caused by the increase of plant biomass also contributed to the decrease of Cd concentrations in plants cultivated in biochar-treated soils [33–35], which suggests that such decrease is not a consequence of biochar application. However, the positive correlations (and no correlations) found for DTPA-extractable Cd in the soil and Cd accumulation in ryegrass plant parts, when the biomass is accounted for, for both biochars, ensure the data reliability obtained in this study (Table 4). The absence of correlation suggests that DTPA-extractable Cd is better correlated with its uptake in plant parts, as shown in the previous study [11].

Table 4. Pearson correlations between DTPA-extractable Cd contents in the soil (x-axis) and Cd accumulation in ryegrass shoots and roots.

Treatment	Shoots	Roots	Shoots + Roots
		Cd accumulation (y-axis)	
SGB	0.60 *	0.50 NS	0.33 **
PLB	0.69 **	0.36 NS	0.47 NS
SGB + PLB	0.28 NS	0.26 NS	0.31 NS

*: $p < 0.05$. **: $p < 0.01$. NS: non-significant ($p > 0.05$). SGB: switchgrass-derived biochar. PLB: poultry litter-derived biochar. SGB + PLB: the two biochar-treated soils.

3.3. Cadmium Transfer Factor

Table 5 shows the Cd transfer factor (TF) from ryegrass roots to shoots. The TF of accumulated Cd was similar to that observed by Antonangelo and Zhang [11] when evaluating the simple Cd uptake TF. Therefore, the TF decreased as the application of both biochars increased (Table 5). A lower TF indicates higher Cd accumulation in roots and lower transference to aboveground plant parts, which lowers the risk to the primary consumer [33,39]. Other studies have also reported the preferential accumulation of Cd in the roots of grasses rather than in shoots, such as ryegrass [31] and rice [32]. In our study, it is evident that the accumulated Cd was higher in the roots of ryegrass than in the shoots even for the control. On average, TF as a function of PLB application was higher than that of SGB, which is consistent with the fact that ryegrass shoots yielded better under the PLB amendment; which was also closely related to the N removal, thus resulted in higher Cd removal aboveground when compared to SGB. The maximum Cd TF was reached when 206 ± 38 kg N ha^{-1} was removed in ryegrass shoots (Figure 3). This is close to a TF of 0.5 obtained between the control and the lowest rate of PLB application rate (0.5%).

Table 5. The transfer factor (TF) of accumulated Cd from ryegrass roots to shoots as a function of biochar (B) application rates (R).

Rate (%)	Transfer Factor (TF)					
	SGB		PLB		Mean	
0	0.66	aA	0.66	aA	0.66	a
0.5	0.25	bB	0.4	bA	0.32	b
1	0.25	bcB	0.35	bA	0.3	b
2	0.28	bB	0.41	bA	0.35	b
4	0.06	cA	0.15	cA	0.11	c
Mean	0.3	B	0.39	A		
R			$p < 0.05$			
B			$p < 0.05$			
R × B			$p = 0.21$			

Different lowercase letters in columns (within biochar rates) and uppercase letters in rows (same biochar rate) are significantly different at $p < 0.05$ (Tukey). SGB: switchgrass-derived biochar. PLB: poultry litter-derived biochar. TF = Cd removal in shoots ÷ Cd accumulated in roots.

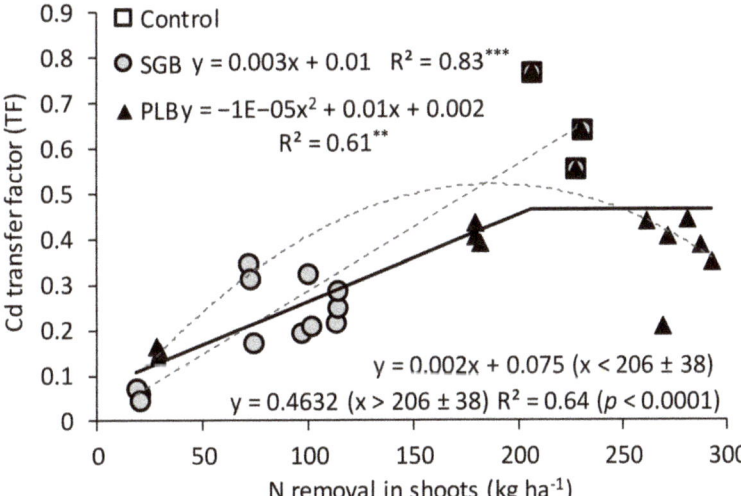

Figure 3. Linear and quadratic regression models (gray dashed lines) for Cd transfer factor and N removal in ryegrass shoots as a function of applications of switchgrass (SGB) and poultry litter-derived biochars (PLB). The black solid line represents a linear regression with a plateau model combining both biochar-treated soils (SGB + PLB). The black dashed vertical line indicates the joint point (plateau point) of N removal where the Cd transfer factor ceases to increase. **: $p < 0.01$. ***: $p < 0.001$.

The linear with plateau model was converged from the whole dataset of measurements and for both biochars combined, which is probably a result of the simple linear regression obtained for SGB and a quadratic regression for PLB (Figure 3). The highest point of which the linear line reaches its maximum in the SGB is close to the maximum point reached with the quadratic line drawn in the PLB before the Cd TF values begin to reduce. Therefore, a maximum point (plateau or joint point), was obtained for both biochars combined thus indicating an optimum N removal in ryegrass shoots in which Cd removal also reached its limit (Figure 3).

4. Discussion

Our research demonstrated that Cd accumulation in ryegrass plant parts is proportionally affected by N uptake and ryegrass yields. Nitrogen fertilization may indeed change the phytoavailability of Cd in the growth medium and its absorption by plants [40]. Additionally, several studies have shown that N fertilization increases Cd concentration in plant tissues [16,41,42]. One of the reasons may rely on a possible synergic effect between nitrate (NO_3^-) and Cd [42–45] while the ratio between NO_3^- and ammonium-NH_4^+ might influence the accumulation of Cd in plants as well [46]. Commonly, NH_4^+ easily undergoes the nitrification process in aerobic environments and forms NO_3^-, which is taken up by plants, and such a process releases H^+ in the soil solution thus reducing its pH and increasing the Cd bioavailability. In this sense, more research is encouraged to test nitrate and ammoniacal sources of nitrogen fertilizers to evaluate their impact on toxic elements removal. Additionally, evaluating if biochars produced from different feedstocks are a better source of NO_3^- or NH_4^+ would contribute to the better understanding of metal remediation practices if biochars can immobilize the metal and at the same time increase the toxic element removal by providing N efficiently while keeping metal concentration below the threshold for animal grazing.

A higher accumulation of Cd was found in the ryegrass roots as compared with the shoots. Since the root is the first plant part in contact with Cd in contaminated soil and its structural component likely accumulates the largest amount of Cd present in the plant tissues [47,48]. This is a physiological strategy in which plants phytostabilize the metal in the roots to prevent toxic elements from reaching the xylem, being transported to the shoots, and damaging the photosynthetic apparatus of plants [46] This prevention of Cd transport to the shoots may occur through the synthesis of chelants or even physical barriers that prevent the Cd movement in the apoplast [49]. Therefore, more attention must be paid to plants metabolism and molecular mechanisms to reveal the direct role of organic amendments, such as biochar, in future phytoremediation studies, as pointed out by Liu et al. [50]. For example, the recent study of Peco et al. [51] has found *Biscutella auriculata* L., a wild herbaceous species that grows on pastureland, as a new Cd-tolerant plant capable of activating efficient metal-sequestering mechanisms in the root surfaces and leaves, and of inducing phytochelatins in both parts, besides stimulating antioxidative defenses in roots.

5. Conclusions

Cadmium removal was the highest at the 1% PLB rate accompanied by the highest ryegrass yield. However, the Cd concentration in grazable forage remained acceptable. The Cd transfer factor from ryegrass roots to shoots increased when up to 206 ± 38 kg N ha^{-1} was removed in ryegrass shoots. Application of up to 1% PLB is a viable option, since it is a practical rate for handling operations requiring less volume of material than SGB.

Author Contributions: Conceptualization, J.A.; methodology, J.A. and H.Z.; validation, J.A. and H.Z.; formal analysis, J.A.; investigation, J.A. and H.Z.; resources, H.Z.; data curation, J.A.; writing—original draft preparation, J.A.; writing—review and editing, J.A. and H.Z.; visualization, J.A.; supervision, H.Z.; project administration, H.Z.; funding acquisition, H.Z. All authors have read and agreed to the published version of the manuscript.

Funding: This work was supported by the Oklahoma Agricultural Experiment Station.

Institutional Review Board Statement: Not applicable.

Informed Consent Statement: Not applicable.

Data Availability Statement: Raw data were generated at an environmentally controlled growth chamber located at CERL (Controlled Environmental Research Lab) Central, Oklahoma State University, main campus. Derived data supporting the findings of this study are available from the corresponding author J.A. on request.

Conflicts of Interest: The authors declare no conflict of interest.

References

1. Zeng, X.; Xu, H.; Lu, J.; Chen, Q.; Li, W.; Wu, L.; Tang, J.; Ma, L. The Immobilization of Soil Cadmium by the Combined Amendment of Bacteria and Hydroxyapatite. *Sci. Rep.* **2020**, *10*, 1–8. [CrossRef] [PubMed]
2. Jacquiod, S.; Cyriaque, V.; Riber, L.; Al-Soud, W.A.; Gillan, D.C.; Wattiez, R.; Sørensen, S.J. Long-term industrial metal contamination unexpectedly shaped diversity and activity response of sediment microbiome. *J. Hazard. Mater.* **2018**, *344*, 299–307. [CrossRef]
3. Martínez-Sánchez, M.J.; Martínez-López, S.; Martínez-Martínez, L.B.; Pérez-Sirvent, C. Importance of the oral arsenic bioaccessibility factor for characterising the risk associated with soil ingestion in a mining-influenced zone. *J. Environ. Manag.* **2013**, *116*, 10–17. [CrossRef]
4. Sinha, S.; Mishra, R.K.; Sinam, G.; Mallick, S.; Gupta, A.K. Comparative Evaluation of Metal Phytoremediation Potential of Trees, Grasses, and Flowering Plants from Tannery-Wastewater-Contaminated Soil in Relation with Physicochemical Properties. *Soil Sediment Contam. Int. J.* **2013**, *22*, 958–983. [CrossRef]
5. He, M.; Shi, H.; Zhao, X.; Yu, Y.; Qu, B. Immobilization of Pb and Cd in Contaminated Soil Using Nano-Crystallite Hydroxyapatite. *Procedia Environ. Sci.* **2013**, *18*, 657–665. [CrossRef]
6. Niu, L.; Yang, F.; Xu, C.; Yang, H.; Liu, W. Status of metal accumulation in farmland soils across China: From distribution to risk assessment. *Environ. Pollut.* **2013**, *176*, 55–62. [CrossRef]
7. Neuberger, J.S.; Hu, S.C.; Drake, K.D.; Jim, R. Potential health impacts of heavy-metal exposure at the Tar Creek Superfund site, Ottawa County, Oklahoma. *Environ. Geochem. Health* **2008**, *31*, 47–59. [CrossRef] [PubMed]
8. Brown, S.; Compton, H.; Basta, N. Field Test of In Situ Soil Amendments at the Tar Creek National Priorities List Superfund Site. *J. Environ. Qual.* **2007**, *36*, 1627–1634. [CrossRef] [PubMed]
9. Beattie, R.E.; Henke, W.; Davis, C.; Mottaleb, M.A.; Campbell, J.H.; Mcaliley, L.R. Quantitative analysis of the extent of heavy-metal contamination in soils near Picher, Oklahoma, within the Tar Creek Superfund Site. *Chemosphere* **2017**, *172*, 89–95. [CrossRef] [PubMed]
10. Gouzie, D. Potential Remediation Methods and Their Applicability to the Tri-State Mining District, USA. In Proceedings of the GSA Annual Meeting, Indianapolis, IN, USA, 4–7 November 2018.
11. Antonangelo, J.A.; Zhang, H. Heavy metal phytoavailability in a contaminated soil of northeastern Oklahoma as affected by biochar amendment. *Environ. Sci. Pollut. Res.* **2019**, *26*, 33582–33593. [CrossRef] [PubMed]
12. Ji, S.; Gao, L.; Chen, W.; Su, J.; Shen, Y. Urea application enhances cadmium uptake and accumulation in Italian ryegrass. *Environ. Sci. Pollut. Res.* **2020**, *27*, 34421–34433. [CrossRef] [PubMed]
13. Singh, J.P.; Singh, B.; Karwasra, S.P.S. Yield and uptake response of lettuce to cadmium as influenced by nitrogen application. *Fertil. Res.* **1988**, *18*, 49–56. [CrossRef]
14. Zhang, R.-R.; Liu, Y.; Xue, W.-L.; Chen, R.-X.; Du, S.-T.; Jin, C.-W. Slow-release nitrogen fertilizers can improve yield and reduce Cd concentration in pakchoi (*Brassica chinensis* L.) grown in Cd-contaminated soil. *Environ. Sci. Pollut. Res.* **2016**, *23*, 25074–25083. [CrossRef]
15. Wei, S.; Ji, D.; Twardowska, I.; Li, Y.; Zhu, J. Effect of different nitrogenous nutrients on the cadmium hyperaccumulation efficiency of *Rorippa globosa* (Turcz.) Thell. *Environ. Sci. Pollut. Res.* **2015**, *22*, 1999–2007. [CrossRef] [PubMed]
16. Liu, W.; Zhang, C.; Hu, P.; Luo, Y.; Wu, L.; Sale, P.; Tang, C. Influence of nitrogen form on the phytoextraction of cadmium by a newly discovered hyperaccumulator *Carpobrotus rossii*. *Environ. Sci. Pollut. Res.* **2016**, *23*, 1246–1253. [CrossRef]
17. Chen, Y.; Liu, M.; Deng, Y.; Zhong, F.; Xu, B.; Hu, L.; Wang, M.; Wang, G. Comparison of ammonium fertilizers, EDTA, and NTA on enhancing the uptake of cadmium by an energy plant, Napier grass (*Pennisetum purpureum* Schumach). *J. Soils Sediments* **2017**, *17*, 2786–2796. [CrossRef]
18. Symanowicz, B.; Kalesa, S.; Jaremko, D.; Niedbała, M. Effect of nitrogen application and year on concentration of Cu, Zn, Ni, Cr, Pb and Cd in herbage of *Galega orientalis* Lam. *Plant Soil Environ.* **2016**, *61*, 11–16. [CrossRef]
19. Benyas, E.; Owens, J.; Seyedalikhani, S.; Robinson, B. Cadmium Uptake by Ryegrass and Ryegrass–Clover Mixtures under Different Liming Rates. *J. Environ. Qual.* **2018**, *47*, 1249–1257. [CrossRef] [PubMed]
20. Liu, L.; Zhang, Q.; Hu, L.; Tang, J.; Xu, L.; Yang, X.; Yong, J.W.H.; Chen, X. Legumes Can Increase Cadmium Contamination in Neighboring Crops. *PLoS ONE* **2012**, *7*, e42944. [CrossRef]
21. Malcolm, B.J.; Moir, J.L.; Cameron, K.C.; Di, H.J.; Edwards, G.R. Influence of plant growth and root architecture of Italian ryegrass (*Lolium multiflorum*) and tall fescue (*Festuca arundinacea*) on N recovery during winter. *Grass Forage Sci.* **2015**, *70*, 600–610. [CrossRef]
22. Mongkhonsin, B.; Nakbanpote, W.; Meesungnoen, O.; Prasad, M.N.V. Adaptive and Tolerance Mechanisms in Herbaceous Plants Exposed to Cadmium. *Cadmium Toxic. Toler. Plants* **2019**, 73–109.
23. Han, S.; Li, X.; Amombo, E.; Fu, J.; Xie, Y. Cadmium Tolerance of Perennial Ryegrass Induced by *Aspergillus aculeatus*. *Front. Microbiol.* **2018**, *9*, 1579. [CrossRef] [PubMed]

24. Antonangelo, J.A.; Zhang, H.; Sun, X.; Kumar, A. Physicochemical properties and morphology of biochars as affected by feedstock sources and pyrolysis temperatures. *Biochar* **2019**, *1*, 325–336. [CrossRef]
25. Church, C.; Spargo, J.; Fishel, S. Strong Acid Extraction Methods for "Total Phosphorus" in Soils: EPA Method 3050B and EPA Method 3051. *Agric. Environ. Lett.* **2017**, *2*, 160037. [CrossRef]
26. Richards, J.R.; Schroder, J.L.; Zhang, H.; Basta, N.T.; Wang, Y.; Payton, M.E. Trace Elements in Benchmark Soils of Oklahoma. *Soil Sci. Soc. Am. J.* **2012**, *76*, 2031–2040. [CrossRef]
27. Wu, J.; Song, Q.; Zhou, J.; Wu, Y.; Liu, X.; Liu, J.; Zhou, L.; Wu, Z.; Wu, W. Cadmium threshold for acidic and multi-metal contaminated soil according to *Oryza sativa* L. Cadmium accumulation: Influential factors and prediction model. *Ecotoxicol. Environ. Saf.* **2021**, *208*, 111420. [CrossRef]
28. Jones, J.B.; Case, V.W. Sampling, Handling, and Analyzing Plant Tissue Samples. *SSSA Book Ser. Soil Test. Plant Anal.* **2018**, *3*, 389–427.
29. Rigby, H.; Smith, S.R. The significance of cadmium entering the human food chain via livestock ingestion from the agricultural use of biosolids, with special reference to the UK. *Environ. Int.* **2020**, *143*, 105800. [CrossRef]
30. Feng, D.; Huang, C.; Xu, W.; Qin, Y.; Li, Y.; Li, T.; Yang, M.; He, Z. Difference of Cadmium Bioaccumulation and Transportation in Two Ryegrass Varieties and the Correlation between Plant Cadmium Concentration and Soil Cadmium Chemical Forms. *Wirel. Pers. Commun.* **2020**, *110*, 291–307. [CrossRef]
31. Jarvis, S.C.; Jones, L.H.P.; Hopper, M.J. Cadmium uptake from solution by plants and its transport from roots to shoots. *Plant Soil* **1976**, *44*, 179–191. [CrossRef]
32. Huang, L.; Li, W.C.; Tam, N.F.Y.; Ye, Z. Effects of root morphology and anatomy on cadmium uptake and translocation in rice (*Oryza sativa* L.). *J. Environ. Sci.* **2019**, *75*, 296–306. [CrossRef]
33. Zhang, G.; Guo, X.; Zhao, Z.; He, Q.; Wang, S.; Zhu, Y.; Yan, Y.; Liu, X.; Sun, K.; Zhao, Y.; et al. Effects of biochars on the availability of heavy metals to ryegrass in an alkaline contaminated soil. *Environ. Pollut.* **2016**, *218*, 513–522. [CrossRef]
34. Lu, K.; Yang, X.; Shen, J.; Robinson, B.; Huang, H.; Liu, D.; Bolan, N.; Pei, J.; Wang, H. Effect of bamboo and rice straw biochars on the bioavailability of Cd, Cu, Pb and Zn to Sedum plumbizincicola. *Agric. Ecosyst. Environ.* **2014**, *191*, 124–132. [CrossRef]
35. Park, J.H.; Choppala, G.K.; Bolan, N.S.; Chung, J.W.; Chuasavathi, T. Biochar reduces the bioavailability and phytotoxicity of heavy metals. *Plant Soil* **2011**, *348*, 439–451. [CrossRef]
36. Al-Wabel, M.I.; Usman, A.R.; El-Naggar, A.H.; Aly, A.A.; Ibrahim, H.M.; Elmaghraby, S.; Al-Omran, A. Conocarpus biochar as a soil amendment for reducing heavy metal availability and uptake by maize plants. *Saudi J. Biol. Sci.* **2015**, *22*, 503–511. [CrossRef]
37. Li, H.; Ye, X.; Geng, Z.; Zhou, H.; Guo, X.; Zhang, Y.; Zhao, H.; Wang, G. The influence of biochar type on long-term stabilization for Cd and Cu in contaminated paddy soils. *J. Hazard. Mater.* **2016**, *304*, 40–48. [CrossRef]
38. Mohamed, I.; Zhang, G.-S.; Li, Z.-G.; Liu, Y.; Chen, F.; Dai, K. Ecological restoration of an acidic Cd contaminated soil using bamboo biochar application. *Ecol. Eng.* **2015**, *84*, 67–76. [CrossRef]
39. Karami, N.; Clemente, R.; Moreno-Jiménez, E.; Lepp, N.W.; Beesley, L. Efficiency of green waste compost and biochar soil amendments for reducing lead and copper mobility and uptake to ryegrass. *J. Hazard. Mater.* **2011**, *191*, 41–48. [CrossRef] [PubMed]
40. Valdez-González, J.C.; López-Chuken, U.J.; Guzmán-Mar, J.L.; Flores-Banda, F.; Hernández-Ramírez, A.; Hinojosa-Reyes, L. Saline irrigation and Zn amendment effect on Cd phytoavailability to Swiss chard (*Beta vulgaris* L.) grown on a long-term amended agricultural soil: A human risk assessment. *Environ. Sci. Pollut. Res.* **2014**, *21*, 5909–5916. [CrossRef]
41. Bauddh, K.; Singh, R.P. Effects of organic and inorganic amendments on bio-accumulation and partitioning of Cd in Brassica juncea and *Ricinus communis*. *Ecol. Eng.* **2015**, *74*, 93–100. [CrossRef]
42. Yang, Y.; Xiong, J.; Chen, R.; Fu, G.; Chen, T.; Tao, L. Excessive nitrate enhances cadmium (Cd) uptake by up-regulating the expression of OsIRT1 in rice (*Oryza sativa*). *Environ. Exp. Bot.* **2016**, *122*, 141–149. [CrossRef]
43. Jalloh, M.A.; Chen, J.; Zhen, F.; Zhang, G. Effect of different N fertilizer forms on antioxidant capacity and grain yield of rice growing under Cd stress. *J. Hazard. Mater.* **2009**, *162*, 1081–1085. [CrossRef]
44. Luo, B.F.; Du, S.T.; Lu, K.X.; Liu, W.J.; Lin, X.Y.; Jin, C.W. Iron uptake system mediates nitrate-facilitated cadmium accumulation in tomato (*Solanum lycopersicum*) plants. *J. Exp. Bot.* **2012**, *63*, 3127–3136. [CrossRef]
45. Hu, J.; Wu, S.; Wu, F.; Leung, H.M.; Lin, X.; Wong, M.H. Arbuscular mycorrhizal fungi enhance both absorption and stabilization of Cd by *Alfred stonecrop* (*Sedum alfredii* Hance) and perennial ryegrass (*Lolium perenne* L.) in a Cd-contaminated acidic soil. *Chemosphere* **2013**, *93*, 1359–1365. [CrossRef]
46. Nogueirol, R.C.; Monteiro, F.A.; Junior, J.C.D.S.; Azevedo, R.A. NO3−/NH4+ proportions affect cadmium bioaccumulation and tolerance of tomato. *Environ. Sci. Pollut. Res.* **2018**, *25*, 13916–13928. [CrossRef] [PubMed]
47. Gallego, S.M.; Pena, L.B.; Barcia, R.A.; Azpilicueta, C.E.; Iannone, M.F.; Rosales, E.P.; Zawoznik, M.S.; Groppa, M.D.; Benavides, M.P. Unravelling cadmium toxicity and tolerance in plants: Insight into regulatory mechanisms. *Environ. Exp. Bot.* **2012**, *83*, 33–46. [CrossRef]
48. Alves, L.R.; Monteiro, C.C.; Carvalho, R.F.; Ribeiro, P.C.; Tezotto, T.; Azevedo, R.A.; Gratão, P.L. Cadmium stress related to root-to-shoot communication depends on ethylene and auxin in tomato plants. *Environ. Exp. Bot.* **2017**, *134*, 102–115. [CrossRef]
49. Lux, A.; Martinka, M.; Vaculik, M.; White, P.J. Root responses to cadmium in the rhizosphere: A review. *J. Exp. Bot.* **2010**, *62*, 21–37. [CrossRef]
50. Liu, M.; Zhao, Z.; Wang, L.; Xiao, Y. Influences of rice straw biochar and organic manure on forage soybean nutrient and Cd uptake. *Int. J. Phytoremediat.* **2020**, *23*, 53–63. [CrossRef]
51. Peco, J.D.; Campos, J.A.; Romero-Puertas, M.C.; Olmedilla, A.; Higueras, P.; Sandalio, M.L. Characterization of mechanisms involved in tolerance and accumulation of Cd in *Biscutella auriculata* L. *Ecotoxicol. Environ. Saf.* **2020**, *201*, 110784. [CrossRef]

Article

Alleviation of Cadmium Adverse Effects by Improving Nutrients Uptake in Bitter Gourd through Cadmium Tolerant Rhizobacteria

Muhammad Zafar-ul-Hye [1], Muhammad Naeem [1], Subhan Danish [1,*], Shah Fahad [2,3,*], Rahul Datta [4,*], Mazhar Abbas [5], Ashfaq Ahmad Rahi [6], Martin Brtnicky [4,7,8], Jiří Holátko [8], Zahid Hassan Tarar [9] and Muhammad Nasir [10]

1. Department of Soil Science, Faculty of Agricultural Sciences and Technology, Bahauddin Zakariya University, Multan 60000, Pakistan; zafarulhyegondal@yahoo.com (M.Z.-u.-H.); ch.naeem276@gmail.com (M.N.)
2. Department of Agronomy, The University of Haripur, Haripur 22620, Pakistan
3. College of Plant Sciences and Technology, Huazhong Agriculture University, Wuhan 430070, China
4. Department of Agrochemistry, Soil Science, Microbiology and Plant Nutrition, Faculty of AgriSciences, Mendel University in Brno, 61300 Brno, Czech Republic; martin.brtnicky@mendelu.cz
5. Institute of Horticultural Sciences, Faculty of Agriculture, University of Agriculture, Faisalabad 38000, Pakistan; rmazhar@hotmail.com
6. Pesticide Quality Control Laboratory, Multan 60000, Pakistan; rahisenior2005@gmail.com
7. Institute of Chemistry and Technology of Environmental Protection, Brno University of Technology, Faculty of Chemistry, Purkynova 118, 62100 Brno, Czech Republic
8. Department of Geology and Pedology, Faculty of Forestry and Wood Technology, Mendel University in Brno, 61300 Brno, Czech Republic; jiri.holatko@centrum.cz
9. Soil and Water Testing Laboratory, Mandi Bahauddin 50400, Pakistan; zahidtarar123@yahoo.com
10. Soil and Water Testing Laboratory for Research, Multan 60000, Pakistan; swt_mltn@yahoo.com
* Correspondence: sd96850@gmail.com (S.D.); shah.fahad@mail.hzau.edu.cn (S.F.); rahulmedcure@gmail.com (R.D.)

Received: 16 May 2020; Accepted: 22 July 2020; Published: 26 July 2020

Abstract: Cadmium is acute toxicity inducing heavy metal that significantly decreases the yield of crops. Due to high water solubility, it reaches the plant tissue and disturbs the uptake of macronutrients. Low uptake of nutrients in the presence of cadmium is a well-documented fact due to its antagonistic relationship with those nutrients, i.e., potassium. Furthermore, cadmium stressed plant produced a higher amount of endogenous stress ethylene, which induced negative effects on yield. However, inoculation of 1-amino cyclopropane-1-carboxylate deaminase (ACCD), producing plant growth promoting rhizobacteria (PGPR), can catabolize this stress ethylene and immobilized heavy metals to mitigate cadmium adverse effects. We conducted a study to examine the influence of ACCD PGPR on nutrients uptake and yield of bitter gourd under cadmium toxicity. Cadmium tolerant PGPRs, i.e., *Stenotrophomonas maltophilia* and *Agrobacterium fabrum* were inoculated solely and in combination with recommended nitrogen, phosphorus, and potassium fertilizers (RNPKF) applied under different concentration of soil cadmium (2 and 5 mg kg^{-1} soil). Results showed that *A. fabrum* with RNPKF showed significant positive response towards an increase in the number of bitter gourds per plant (34% and 68%), fruit length (19% and 29%), bitter gourd yield (26.5% and 21.1%), N (48% and 56%), and K (72% and 55%) concentration from the control at different concentrations of soil cadmium (2 and 5 mg kg^{-1} soil), respectively. In conclusion, we suggest that *A. fabrum* with RNPKF can more efficaciously enhance N, K, and yield of bitter gourd under cadmium toxicity.

Keywords: ACC deaminase; heavy metal stress; PGPR; fertilizers; nutrients; yield

1. Introduction

High use of pesticides, inorganic fertilizers, and untreated sewage water has significantly contributed to the buildup of heavy metals in agricultural soils [1,2]. These heavy metals become part of soil at the exchange site and remain readily available for plants. Rapid industrialization and anthropogenic activities are also allied factors responsible for the accumulation of toxic metals beyond their threshold limit in cultivatable lands [3–6]. Among different heavy metals, cadmium (Cd) is an acute toxin due to its high resistance time, i.e., >1000 years and water solubility [7]. Presence of cadmium below 0.5 mg kg^{-1} soil is considered a safe limit, but depending upon parent material, it can be accumulated up to 3.0 mg kg^{-1} soil [8]. Being a part of phosphate fertilizers (up to 4.4 mg kg^{-1}), it is easily taken up by crops as Cd-supplement [9,10].

Cadmium causes cardiovascular, respiratory, cancer, and renal, skeletal system in humans when taken up beyond the threshold limit [11,12]. The high concentration of Cd in plant tissues disturbs nutrient uptake and creates water imbalance that results in poor photosynthesis [13]. It also causes lipid membrane instability, alteration in membranes permeability, and chlorosis in plants [14,15]. Due to its divalent nature, it competes with divalent essential nutrients, i.e., P, Ca, Mg, and decreases their uptake in plants [16–18]. Bioavailability of K is also affected when Cd is present in higher concentration in soil [19]. Heavy metal toxicity and physiochemical properties soil depend on the land use [20–22]. Different crops, the biological adsorption factor (mg Cd/kg plant ash, mg Cd/kg soil) based on Cd content in plant ash, is different, i.e., winter wheat grains (5.97) and straw (2.50), barley grains (4.06) and straw (2.50), sugar beet roots (4.63) and tops (1.41), pea beans (3.22) and straw (0.88), corn grains (8.75) and straw (2.53), soya beans (4.31) and straw (4.63), sunflower seeds (10.8) and stem (4.28) [23]. Moreover, biosynthesis of endogenous stress ethylene under Cd toxicity plays a notorious role that aids in poor root growth [24–26].

In addition, ethylene (C_2H_4) is a plant-signaling molecule. It is involved in seed germination flower senescence, root elongation, fruit ripening, and leaf abscission. Mostly ethylene is synthesized in a two-step process, i.e., (1) enzymatic conversion of S-adenosyl methionine (SAM) to 1-amino cyclopropane-1-carboxylic acid (ACC); (2) conversion of ACC to ethylene, which is catalyzed by ACC-oxidase [27]. However, synthesis of endogenous ethylene level is significantly enhanced upon exposure of plants to abiotic stresses, including low soil fertility [28,29]. This endogenous stress ethylene negatively affects gas exchange attributes, nutrients and water uptake, and yield of different crops under any stress conditions [30,31].

To overcome these problems, inoculation of ACC deaminase producing plant growth promoting rhizobacteria (PGPR), could be an efficacious and nature friendly technique [32–36]. Certain PGPRs can improve growth attributes of crops under heavy metals toxic conditions by secretions of ACC deaminase, siderophores, indole acetic acid, gibberellins, and better availability of water and nutrients [37–40]. Enzyme ACC deaminase cleaves stress ethylene into intermediate compounds; thus, decreases the stress generating factors in plants [41,42].

Among different crop plants, bitter gourd is a rich source of vitamins, carbohydrates, and proteins [43,44]. As compared to cucumber and tomato, one cup of bitter gourd juice (94 g) has 93% reference daily intake (RDI) of vitamin C [45]. It is cultivated in Pakistan (6107 hectares), with an annual production of 57,190 tons [46]. However, the yield of bitter gourd is negatively affected when cultivated in Cd pollution. As improvement in N, P, and K can mitigate the stress of Cd toxicity in plants [3], which is why the current study was conducted to explore the efficacy of ACC deaminase producing PGPR with recommended NPK fertilizers (RNPKF) on bitter gourd nutrients uptake and yield under Cd toxicity.

The present study aimed to explore (1) effectiveness of rhizobacteria in the improvement of nutrients uptake; (2) effect of nutrients on bitter gourd yield under cadmium-induced stress; (3) correlation of inorganic fertilizer with rhizobacteria on yield and nutrients attributes of bitter gourd under Cd stress. We hypothesized that ACC deaminase-producing PGPRs could improve nutrient uptake and alleviate adverse effects of Cd in bitter gourd for yield improvement.

2. Materials and Methods

2.1. Experimental Site and Treatments

A pot experiment was conducted in the Department of Soil Science research area, Bahauddin Zakariya University, Multan, Pakistan. The soil was characterized as dark brown and saline with JAKHAR soil series [42]. Six treatments were applied in four replication by following two factorial completely randomized designs (CRDs). The treatments were control (without NPK or bacterial strains), recommended NPK fertilizers (RNPKF), *Stenotrophomonas maltophilia*, *Agrobacterium fabrum*, RNPKF + *S. maltophilia*, and RNPKF + *A. fabrum*. All treatments were added in the soil at 2 and 5 mg Cd kg^{-1} soil. Artificial toxicity of Cd was developed by using analytical grade salt of $CdCl_2$ [25]. As per treatment plan, two levels of Cd were maintained, i.e., 2 and 5 ppm (mg kg^{-1} soil), keeping in mind the Cd concentration of pre-experimental soil. Rhizobacteria were inoculated at the time of sowing. However, required fertilizers were applied at the time of pot preparation.

2.2. Collection of Bacterial Strains and Broth

ACC deaminase producing PGPRs *S. maltophilia* (ACC deaminase activity = 71.78 μmol α-ketobutyrate mg^{-1} protein h^{-1}) and *A. fabrum* (ACC deaminase activity = 432.6 μmol α-ketobutyrate mg^{-1} protein h^{-1}) were taken from the Laboratory of Soil Microbiology, Department of Soil Science. Both PGPRs were documented Cd tolerant previously, i.e., survive over 5.0 mg Cd kg^{-1} soil toxicity [25]. For seeds inoculation, Dworkin and Foster (DF) media without agar was used for inoculum preparation [47]. Loop of each rhizobacteria was taken in the sterilized flask and incubated at 25 ± 3 °C and 100 rpm for 48 h. After that, broth optical density (OD) was measured by spectrophotometer (540 nm wavelength). Finally, dilution was made with sterilized distilled water to achieve 0.45 nm OD, to achieve a uniform population of 10^7–10^8 colony forming units (CFU) mL^{-1}.

2.3. Seeds Sterilization and Sowing

$HgCl_2$ (0.1%) solution was used for sterilization of seeds. All seeds were placed for 5 min in the solution followed by, three times, washing with sterilized deionized water [48]. Moreover, 1mL respective PGPR inoculum was used for seeds inoculation along with sugar (30% sucrose), peat, and clay (1:1) in 1:2:6 ratio for 100 g seeds. Four inoculated seeds were sown in each pot. Sowing of bitter gourd seeds was done by hand. After 20 days of sowing, only three healthy seedlings were maintained in each pot by thinning.

2.4. Irrigation and Fertilizer Application

In pots, 65% field capacity was maintained on a weight basis during the experiment. To fulfil the requirement of crop nutrients (187N, 75P, and 225K kg ha^{-1}) urea, K_2HPO_4 and K_2SO_4 were applied.

2.5. Harvesting and Samples Analyses

Harvesting was done at the time of fruit maturity. Samples were digested for the determination of biochemical attributes. The number of bitter gourds was counted manually. For fruit length, standard measuring tape was used. For determination of yield per plant, fruits were collected and weighed on the analytical balance. With the help of diacid mixture nitric acid and perchloric acid (2:1 ratio), the tissues of the plant were digested for P and K analyses [49]. Phosphorus in the samples was determined by using ammonium molybdate and ammonium metavanadate yellow color method [50].

However, for analyses of K in samples, the digested solution was run on flame photometer [51]. For determination of nitrogen, samples were digested in concentrated H_2SO_4, and digestion mixture (K_2SO_4 (100):$CuSO_4.5H_2O$ (10):$FeSO_4$ (1)). Distillation was performed in Kjeldahl distillation apparatus, using boric acid as a collector [52].

2.6. Statistical Analyses

One-way ANOVA was used to assess the effects of treatments. Two factorial ANOVA was conducted separately to compare PGPRs and RNPK interaction under different levels of Cd. Treatment comparison was computed at $p \leq 0.05$ by Tukey's Test.

3. Results

3.1. Number of Bitter Gourds per Plant

Effects of PGPRs and RNPKF under different levels of Cd were significant ($p \leq 0.05$) on the number of bitter gourds per plants (BDP). Inoculation of PGPRs and RNPKF have significant main and interactive effects on BDP at 2 and 5 mg kg^{-1} soil. Application of RNPKF + S. *maltophilia*, RNPKF + A. *fabrum*, RNPKF, S. *maltophilia* and A. *fabrum* showed significant positive effect over control at 2 and 5 mg Cd kg^{-1} soil for BDP (Figure 1). Interaction between PGPRs and RNPKF at 2 mg Cd kg^{-1} soil (Figure 2A) and 5 mg Cd kg^{-1} soil (Figure 2B) were significant ordinal for BDP (Figure 2B). It was noted that Cd showed non-significant negative correlation (−0.1451; $p = 0.3986$) with BDP. However, PGPR (0.5863; $p = 0.0002$) and RNPKF (0.3237; $p = 0.0541$) showed positive significant correlation with BDP (Figure 3). The maximum increase of 34% and 68% in BDP was observed from control where RNPKF + A. *fabrum* was applied at 2 and 5 mg Cd kg^{-1} soil, respectively.

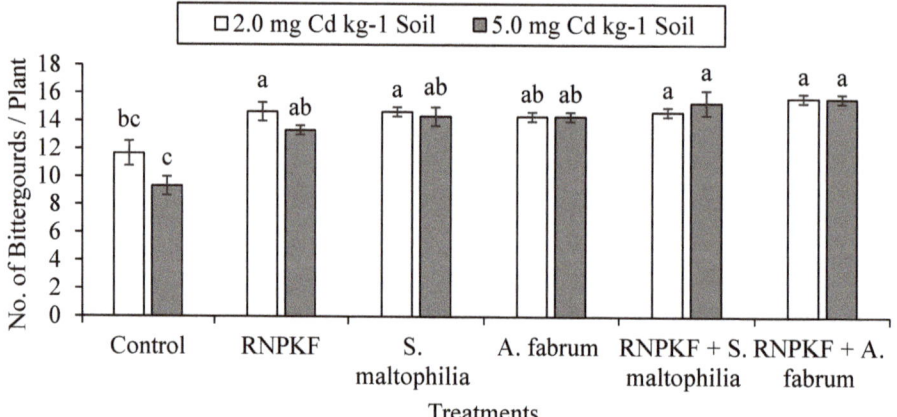

Figure 1. Number of bitter gourds per plant (BDP) treated with plant growth promoting rhizobacteria (PGPRs), recommended NPK fertilizers (RNPKF), and their combination under 2 and 5 mg Cd kg^{-1} soil. Different small letters express significant differences ($p \leq 0.05$).

Estimated Marginal Means of BDP 2 mg Cd kg^{-1} soil **Estimated Marginal Means of BDP 5 mg Cd kg^{-1} soil**

(A) $p < 0.05$ **(B)** $p < 0.05$

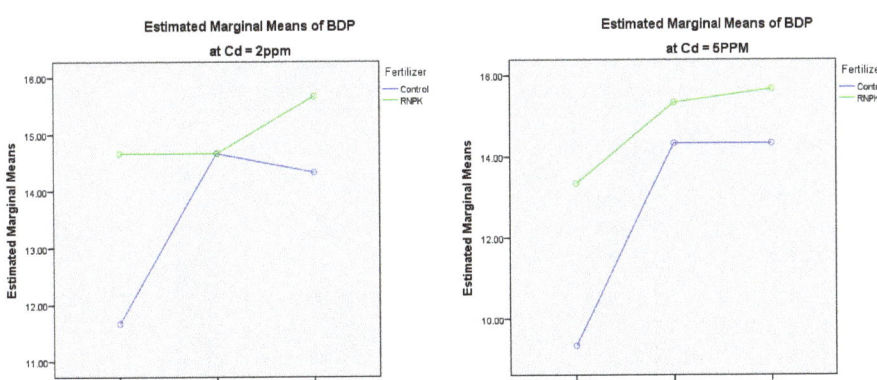

Figure 2. Interaction graph of *S. maltophilia* (NS1), *A. fabrum* (NS2), and RNPKF at 2 (**A**) and 5 mg Cd kg^{-1} soil (**B**) for number of bitter gourd per plant (BDP).

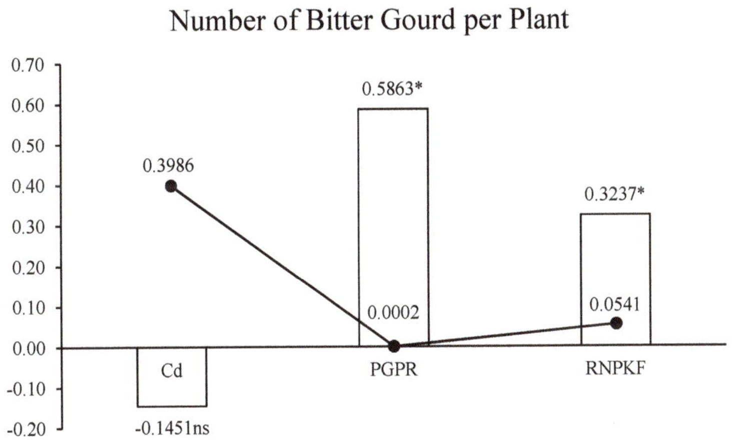

Figure 3. Pearson correlation of Cd, PGPRs, and RNPKF for number of bitter gourd per plant (BDP). * = significant ($p \leq 0.05$); ns = non-significant.

3.2. Bitter Gourd Fruit Length

Effects of PGPRs inoculation and application of RNPKF under various Cd levels were significant ($p \leq 0.05$) on bitter gourd fruit length (FL). Application of RNPKF + *S. maltophilia* and RNPKF + *A. fabrum* were significantly different from control at 2 and 5 mg Cd kg^{-1} soil for FL. It was observed that RNPKF showed a positive significantly better response at 2 mg Cd kg^{-1} soil but remained non-significant at 5 mg Cd kg^{-1} soil over control for FL (Figure 4). Main effects of PGPRs and RNPKF were significant, but their interaction remained non-significant for FL at 2 and 5 mg kg^{-1} soil. Disordinal non-significant interaction was found between PGPRs and RNPKF at 2 mg Cd kg^{-1} soil, but the interaction was non-significant ordinal at 5 mg Cd kg^{-1} soil for FL. Cadmium showed significant but negative correlation (−0.6399; $p = 0.0001$) with FL. Inoculation of PGPRs (0.2239; $p = 0.1893$) gave non-significant positive correlation, while RNPKF (0.3835; $p = 0.021$) showed positive significant correlation with FL (Figure 5). The maximum increase of 19 and 29% in FL was observed from control where RNPKF + *A. fabrum* was applied at 2 and 5 mg Cd kg^{-1} soil, respectively.

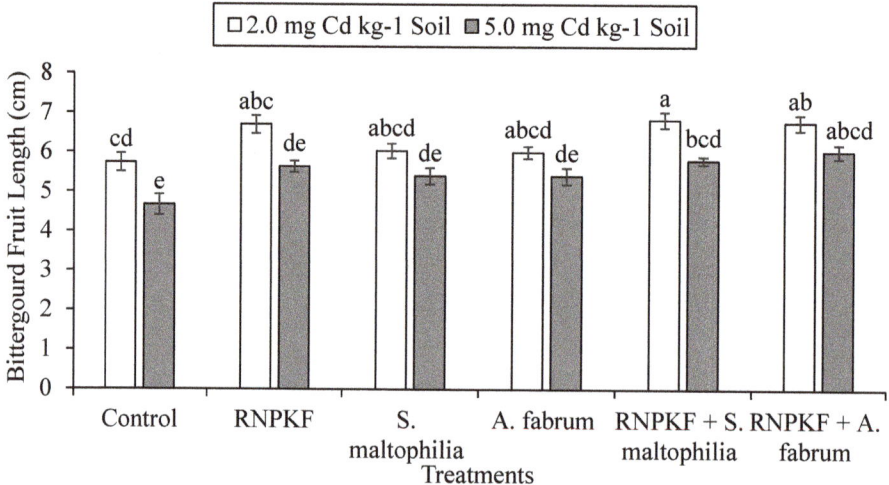

Figure 4. Bitter gourd fruit length (cm) treated with PGPRs, RNPKF, and their combination under 2 and 5 mg Cd kg^{-1} soil. Different small letters express significant differences at $p \leq 0.05$.

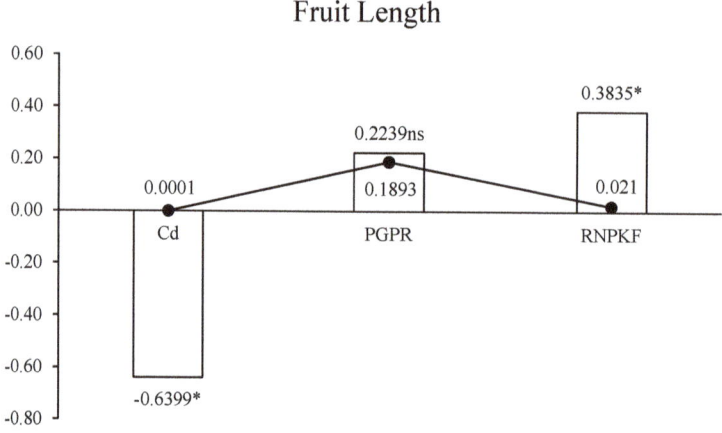

Figure 5. Pearson correlation of Cd, PGPRs, and RNPKF for fruit length (FL). * = significant ($p \leq 0.05$); ns = non-significant.

3.3. Bitter Gourd Yield per Plant

PGPRs *S. maltophilia* and *A. fabrum* and RNPKF under 2 and 5 mg Cd kg^{-1} soil significantly ($p \leq 0.05$) affect bitter gourd yield per plant (YP). At 2 mg Cd kg^{-1} soil, inoculation of PGPRs has significant main and interactive effects on YP. Application of RNPKF has a significant main effect on YP at 5 mg Cd kg^{-1} soil. Treatment RNPKF + *A. fabrum* was significantly different as compared to control at 2 and 5 mg Cd kg^{-1} soil Cd for YP (Figure 6). It was observed that the interaction of PGPRs and RNPKF was significant ordinal at 2 mg Cd kg^{-1} soil (Figure 7A) but non-significant ordinal at 5 mg Cd kg^{-1} soil for YP (Figure 7B). Heavy metal Cd showed significant negative correlation (−0.4385; $p = 0.0075$) with YP. However, PGPRs (0.5035; $p = 0.0017$) and RNPKF (0.3829; $p = 0.0212$) showed positive significant correlation with YP (Figure 8). The maximum increase of 26.5 and 21.1% in YP was observed from control, where RNPKF + *A. fabrum* was applied at 2 and 5 mg Cd kg^{-1} soil, respectively.

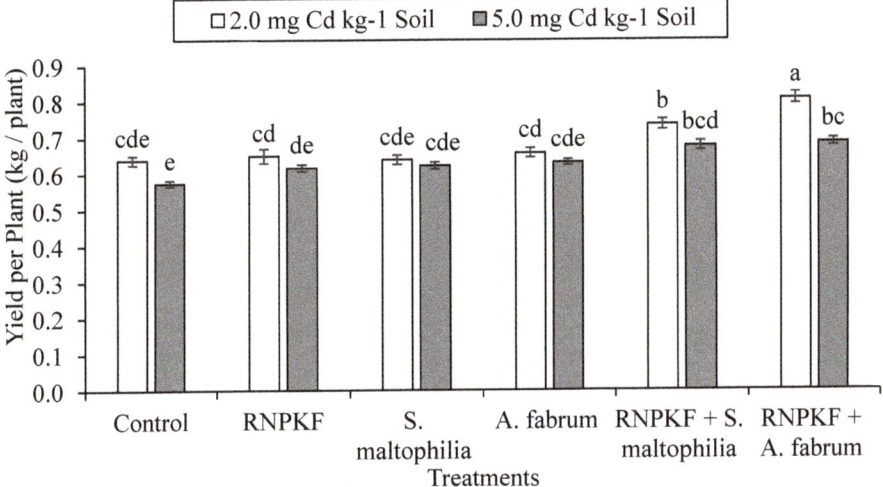

Figure 6. Bitter gourd yield (kg) per plant treated with PGPRs, RNPKF, and their combination under 2 and 5 mg Cd kg^{-1} soil. Different small letters express significant differences at $p \leq 0.05$.

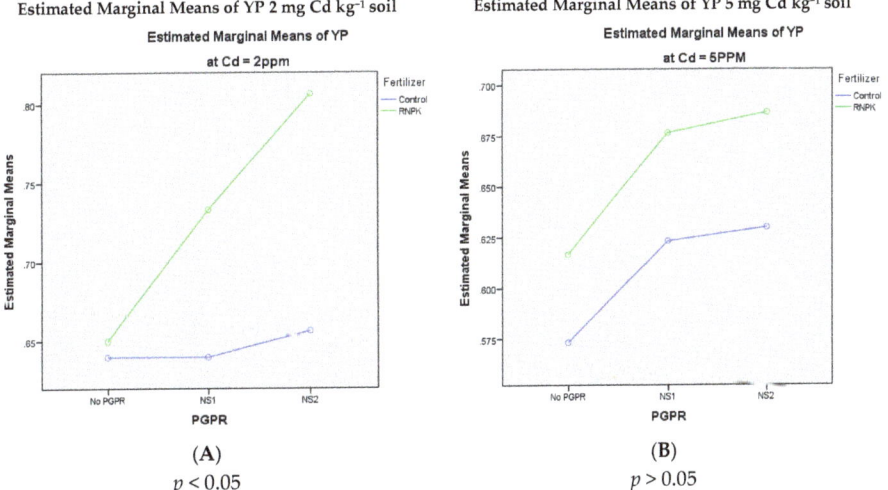

Figure 7. Interaction graph of *S. maltophilla* (NS1), *A. fabrum* (NS2), and RNPKF at 2 (**A**) and 5 mg Cd kg^{-1} soil (**B**) for bitter gourd yield per plant (YP).

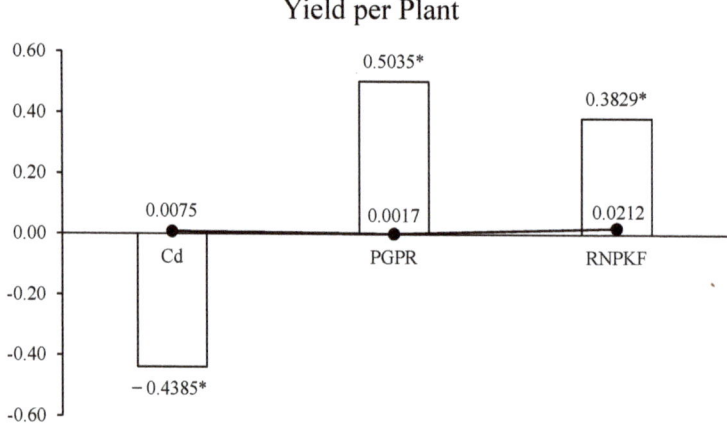

Figure 8. Pearson correlation of Cd, PGPRs, and RNPKF for yield per plant (YP)* = significant ($p \leq 0.05$); ns = non-significant.

3.4. Nitrogen Concentration in Bitter Gourd

PGPRs and RNPKF significantly ($p \leq 0.05$) changed the nitrogen concentration of bitter gourd (NB) under different levels of Cd. Main effects of PGPRs and RNPKF were significant on NB at 2 and 5 mg kg^{-1} soil. However, the interaction of PGPRs and RNPKF was non-significant, ordinal at 2 and 5 mg Cd kg^{-1} soil for NB. It was observed that RNPKF + *S. maltophilia* and RNPKF + *A. fabrum* were significantly different as compared to control at 2 and 5 mg Cd kg^{-1} soil for NB (Figure 9). Heavy metal Cd showed significant negative correlation (-0.4812; $p = 0.0030$) with NB. However, PGPRs (0.4391; $p = 0.0074$) showed significant and RNPKF (0.2041; $p = 0.2324$) showed non-significant positive correlation with NB (Figure 10). The maximum increase of 48 and 56% in NB was observed from control where RNPKF + *S. maltophilia* was applied at 2 and 5 mg Cd kg^{-1} soil, respectively.

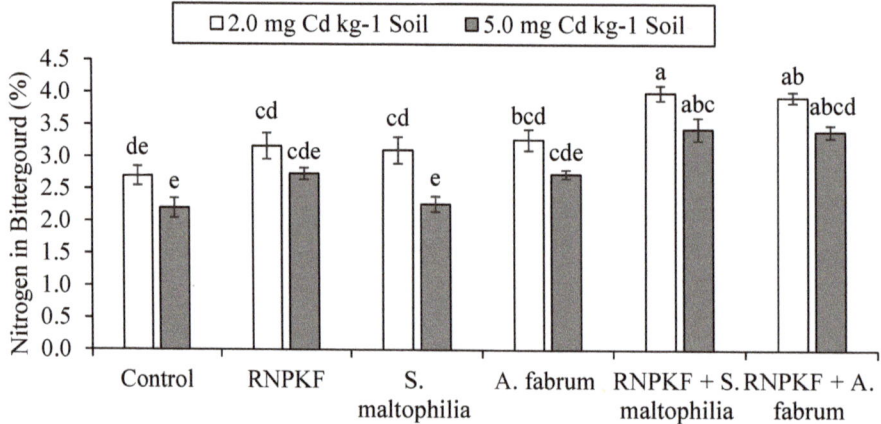

Figure 9. Nitrogen concentration in bitter gourd (%) treated with PGPRs, RNPKF, and their combination under 2 and 5 mg Cd kg^{-1} soil. Different small letters express significant differences ($p \leq 0.05$).

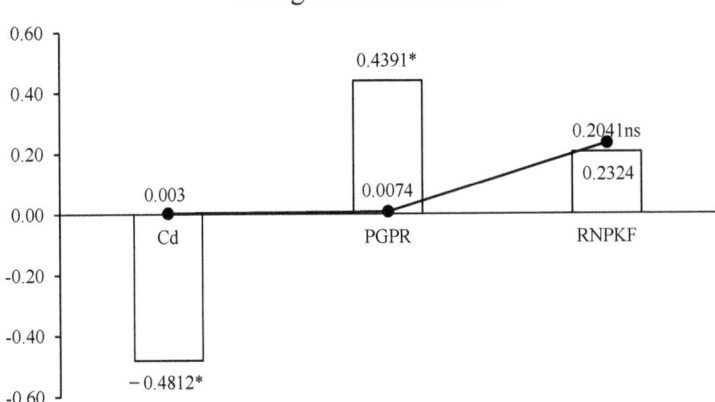

Figure 10. Pearson correlation of Cd, PGPRs, and RNPKF for bitter gourd nitrogen concentration (NB) * = significant ($p \leq 0.05$); ns = non-significant.

3.5. Phosphorus Concentration in Bitter Gourd

Effect of PGPRs and RNPKF under 2 and 5 mg kg^{-1} soil was significant ($p \leq 0.05$) on phosphorus concentration of bitter gourd (PB). Treatments RNPKF, RNPKF + *S. maltophilia*, RNPKF + *A. fabrum*, and RNPKF differed significantly at 5 mg Cd kg^{-1} soil over control for PB (Figure 11). Application of RNPKF and PGPRs showed a significant main effect on PB at 2 mg Cd kg^{-1} soil. At 5 mg Cd kg^{-1} soil, PGPRs and RNPKF have a significant main and interactive effect on PB. Ordinal interaction was found between PGPRs and RNPKF at 2 mg Cd kg^{-1} soil but significant ordinal interaction was observed at 5 mg Cd kg^{-1} soil (Figure 12A,B) for PB. Cadmium showed a significant negative correlation (−0.6614; $p = 0.0001$) with BDP. However, PGPR (0.2537; $p = 0.1953$) showed non-significant and RNPKF (0.4422; $p = 0.0069$) showed significant positive correlation with PB (Figure 13). The maximum increase of 29.5% in PB was observed from control where RNPKF + *A. fabrum* was applied at 2 mg Cd kg^{-1} soil.

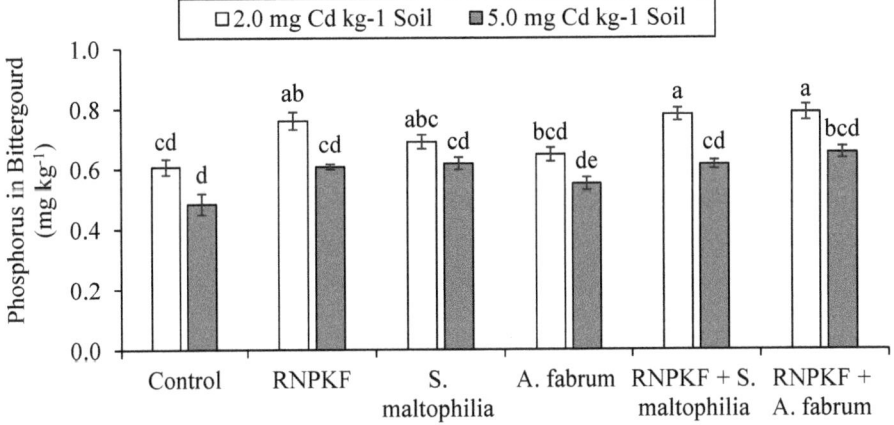

Figure 11. Phosphorus concentration in bitter gourd (%) treated with PGPRs, RNPKF, and their combination under 2 and 5 mg Cd kg^{-1} soil. Different small letters express significant differences ($p \leq 0.05$).

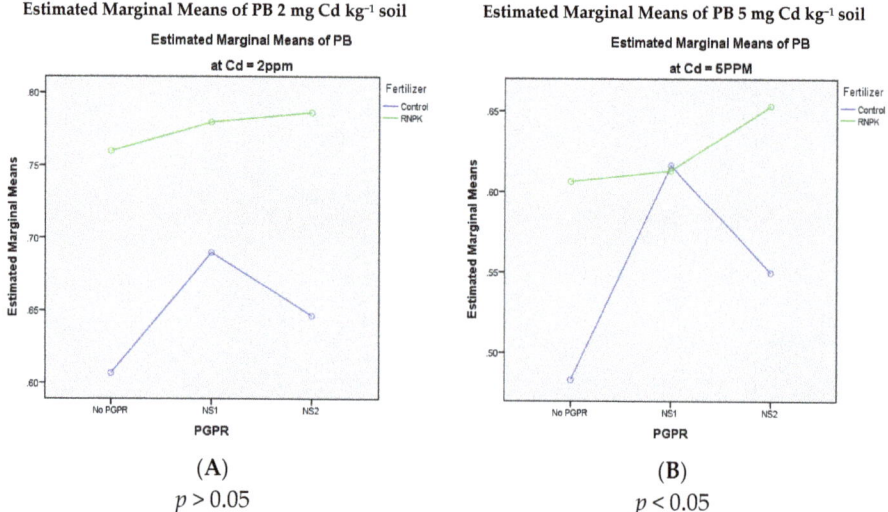

Figure 12. Interaction graphs of *S. maltophilia* (NS1), *A. fabrum* (NS2), and RNPKF at 2 (**A**) and 5 mg Cd kg^{-1} soil (**B**) for phosphorus concentration in bitter gourd (PB).

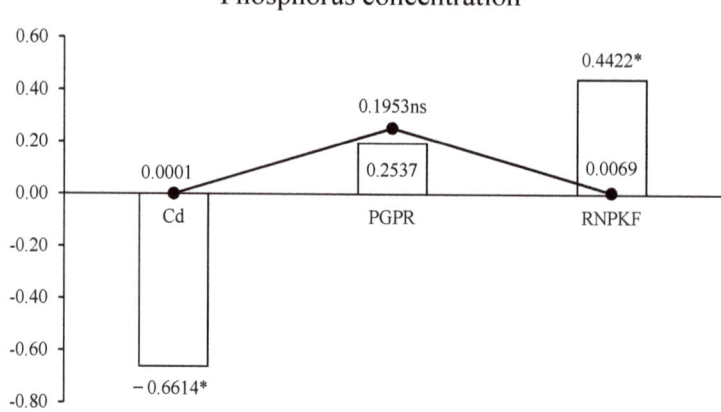

Figure 13. Pearson correlation of Cd, PGPRs, and RNPKF for bitter gourd phosphorus concentration (NB). * = significant ($p \leq 0.05$); ns = non-significant.

3.6. Potassium Concentration in Bitter Gourd

Influence of PGPRs and RNPKF 2 and 5 mg kg^{-1} soil was significant ($p \leq 0.05$) on potassium concentration of bitter gourd (KB). It was also observed that RNPKF + *S. maltophilia* and RNPKF + *A. fabrum* differed significantly for KB at 5 mg Cd kg^{-1} soil (Figure 14). Both PGPRs and RNPKF have a significant main effect on KB at 2 and 5 mg Cd kg^{-1} soil. Disordinal non-significant interaction was found between PGPRs and RNPKF at 2 mg Cd kg^{-1} soil and but ordinal interaction was observed at 5 mg Cd kg^{-1} soil for KB. Different levels of Cd showed significant negative correlation (−0.4904; $p = 0.0024$) with KB. However, PGPR (0.5516; $p = 0.0005$) and RNPKF (0.3840; $p = 0.0208$) showed significant positive correlation with KB (Figure 15). Application of RNPKF + *S. maltophilia*, RNPKF + *A. fabrum*, RNPKF, *S. maltophilia* and *A. fabrum* were significantly different as compared to control at 2 mg Cd kg^{-1} soil for KB. The maximum increase of 72 and 55% in KB was observed from control where RNPKF + *A. fabrum* was applied at 2 and 5 mg Cd kg^{-1} soil, respectively.

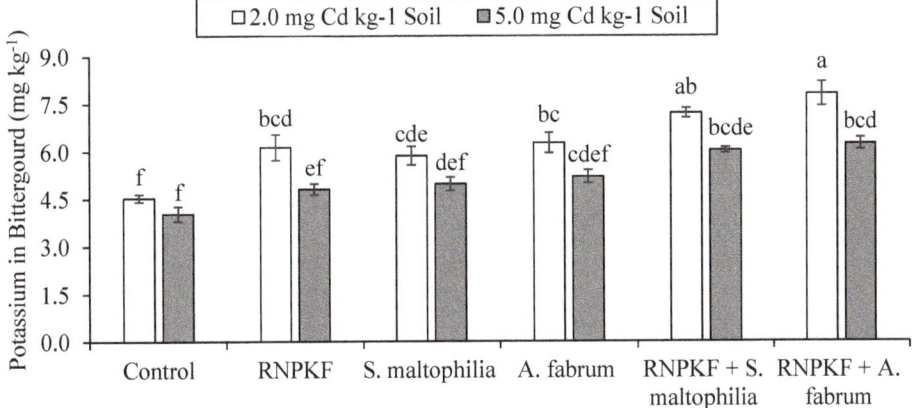

Figure 14. Potassium concentration in bitter gourd (%) treated with PGPRs, RNPKF, and their combination under 2 and 5 mg Cd kg^{-1} soil. Different small letters express significant differences ($p \leq 0.05$).

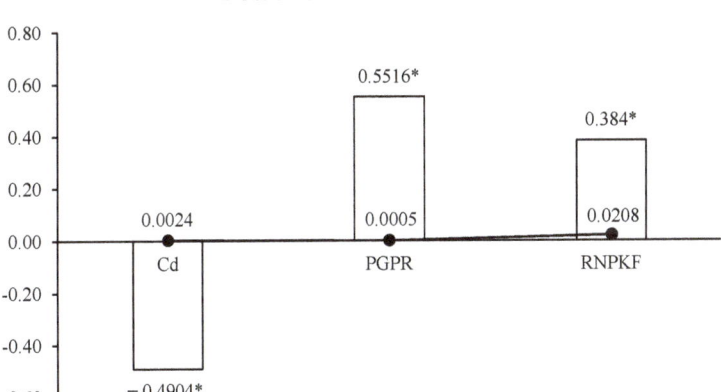

Figure 15. Pearson correlation of Cd, PGPRs, and RNPKF for bitter gourd potassium concentration (NB). * = significant ($p \leq 0.05$); ns = non-significant.

4. Discussion

A significant decrease in fruit length, fresh weight, and yield per plant of bitter gourd were observed in control at 5 mg Cd kg^{-1} soil. Low uptake of N, P, and K in bitter gourd under Cd toxicity might be a major factor for reduction in yield, fruit length, and fresh weight. Higher biosynthesis of stress ethylene might be an allied factor responsible for a significant decline in yield of bitter gourd under Cd stress. According to Sanita di Toppi and Gabbrielli [7], accumulation of Cd beyond safe limit disturbed the nutrients homeostasis which played an imperative role in reduction of root and shoot elongation. Cadmium in plants also competes with divalent nutrients ions and decreases their uptake in plants [16]. Under Cd toxicity, transmembrane carriers in roots become unable to distinguish between non-essential Cd and essential divalent nutrients during their uptake [53,54]. Glick et al. [55] also documented that biosynthesis of endogenous stress ethylene under abiotic stress conditions, negatively affects the productivity of the crop. Toxicity of heavy metals causes abnormal division of cell thus induced chromosomal aberration in plants [56]. This resulted in a decrease of protochlorophyllide reductase activity. Such disturbance in plants also induced chlorosis in leaves [57].

Furthermore, Matile et al. [58] suggested the decomposition of lipids in cell wall when ethylene concentration is increased. They argued that ethylene when contact with chlorophyllase (chlase) gene it degrades chlorophyll caused in chlorosis. Furthermore, application of RNPKF + *A. fabrum* differed significantly better from the sole application of control for improvement in N, P and K. The improvement in N, P, and K mitigate the adverse impacts of Cd in bitter gourd. Panković et al. [59] observed that improvement in N uptake of sunflower allevaints the inhibitory influences of Cd [22,23,27,28,32]. Higher N facilitates in activity of Rubisco by an increase in soluble protein contents. Application of N in NH_4 form is efficacious in decreasing the Cd uptake due to antagonistic relationship [60]. Findings of the current experiment also support the above argument. Better N in bitter gourd was observed where yield was improved over control even under Cd toxicity. Under Cd stress, plants start producing N metabolites, i.e., proline that causes phytochelation and decreases the intake of Cd [61]. Application of phosphorus neutralizes the adverse impacts of Cd and improve the yield of crops [62]. Improvement of P uptake in plants enhances the synthesis of glutathione that prevents membrane damages caused by Cd [63]. Balance K concentration decreases the generation of reactive oxidative species (ROS) and inhibits the NADPH oxidase [64]. Moreover, less generation of stress ethylene by inoculation of *A. fabrum* and RNPKF + *A. fabrum* might be another major factor responsible for the enhancement in bitter gourd growth and yield in the current study. Both PGPRs were capable to produce ACC deaminase, which cleaves ethylene into intermediate compounds. Similar kinds of results were also documented by many scientists [25,26,30,31]. Glick et al. [44] proposed that enzyme ACC deaminase break ethylene into α-ketobutyrate and ammonia [65,66]. Accumulated ethylene in roots moved towards rhizosphere; thus, ethylene becomes low in plant roots, and stress is alleviated. Similarly, Tripathi et al. [67] reported growth hormones, indole acetic acid, improved the root elongation for better uptake of nutrients [24].

5. Conclusions

It is concluded that PGPR, *A. fabrum* has more potential over *S. maltophilia* to alleviate Cd induced stress in bitter gourd. Inoculation of *A. fabrum* with RNPKF is an efficacious approach to improve N, P, and K concentration in bitter gourd. The combined use of RNPKF and *A. fabrum* can increase the number of bitter gourds per plant, bitter gourd fruit length, and yield per plant by alleviating 5 mg Cd kg^{-1} soil induced toxicity. However, more investigations are suggested at field level to declare *A. fabrum* + RNPKF as an efficacious technique to mitigate Cd stress in bitter gourd.

Author Contributions: M.Z.-u.-H. and S.D. designed and supervised the experiment and wrote the manuscript; M.N. (Muhammad Naeem) conducted research, collected data; S.D., M.B., J.H., and R.D. wrote the manuscript and conducted statistical analyses; S.F., M.A., A.A.R., Z.H.T., and M.N. (Muhammad Nasir) assisted in the preparation of manuscript and reviewed manuscript. All authors have read and agreed to the published version of the manuscript.

Funding: This research received no external funding.

Acknowledgments: This research article is part of Muhammad Naeem Thesis for the award of M.Sc. Hons. Agriculture (Soil Science) Degree.

Conflicts of Interest: The authors declare no conflict of interest.

References

1. Fu, F.; Wang, Q. Removal of heavy metal ions from wastewaters: A review. *J. Environ. Manag.* **2011**, *92*, 407–418. [CrossRef] [PubMed]
2. Meena, R.S.; Kumar, S.; Datta, R.; Lal, R.; Vijayakumar, V.; Brtnicky, M.; Sharma, M.P.; Yadav, G.S.; Jhariya, M.K.; Jangir, C.K. Impact of Agrochemicals on Soil Microbiota and Management: A Review. *Land* **2020**, *9*, 34. [CrossRef]

3. Nazar, R. Cadmium Toxicity in Plants and Role of Mineral Nutrients in Its Alleviation. *Am. J. Plant Sci.* **2012**, *3*, 1476–1489. [CrossRef]
4. Lazar, V.; Cernat, R.; Balotescu, C.; Cotar, A.; Coipan, E.; Cojocaru, C. Correlation between Multiple Antibiotic Resistance and Heavy-Metal Tolerance among some E.coli Strains Isolated from Polluted Waters. *Bacteriol. Virusol. Parazitol. Epidemiol. (Buchar. Rom. 1990)* **2002**, *47*, 155–160.
5. Molaei, A.; Lakzian, A.; Haghnia, G.; Astaraei, A.; Rasouli-Sadaghiani, M.; Ceccherini, M.T.; Datta, R. Assessment of some cultural experimental methods to study the effects of antibiotics on microbial activities in a soil: An incubation study. *PLoS ONE* **2017**, *12*, e0180663. [CrossRef] [PubMed]
6. Molaei, A.; Lakzian, A.; Datta, R.; Haghnia, G.; Astaraei, A.; Rasouli-Sadaghiani, M.; Ceccherini, M.T. Impact of chlortetracycline and sulfapyridine antibiotics on soil enzyme activities. *Int. Agrophys.* **2017**, *31*, 499–505. [CrossRef]
7. Sanita di Toppi, L.; Gabbrielli, R. Response to cadmium in higher plants. *Environ. Exp. Bot.* **1999**, *41*, 105–130. [CrossRef]
8. Vahter, M.; Berglund, M.; Slorach, S.; Friberg, L.; Sarić, M.; Zheng, X.; Fujita, M. Methods for integrated exposure monitoring of lead and cadmium. *Environ. Res.* **1991**, *56*, 78–89. [CrossRef]
9. Kabata-Pendias, A.; Pendias, H. *Trace Elements in Soils and Plants*, 2nd ed.; CRC Press: Boca Raton, FL, USA, 2001; p. 331.
10. Muramoto, S.; Aoyama, I. Effects of fertilizers on the vicissitude of cadmium in rice plant. *J. Environ. Sci. Health Part A Environ. Sci. Eng. Toxicol.* **1990**, *25*, 629–636. [CrossRef]
11. Radwan, M.A.; Salama, A.K. Market basket survey for some heavy metals in Egyptian fruits and vegetables. *Food Chem. Toxicol.* **2006**, *44*, 1273–1278. [CrossRef]
12. Steenland, K.; Boffetta, P. Lead and cancer in humans: Where are we now? *Am. J. Ind. Med.* **2000**, *38*, 295–299. [CrossRef]
13. Hossain, M.A.; Hasanuzzaman, M.; Fujita, M. Up-regulation of antioxidant and glyoxalase systems by exogenous glycinebetaine and proline in mung bean confer tolerance to cadmium stress. *Physiol. Mol. Biol. Plants* **2010**, *16*, 259–272. [CrossRef] [PubMed]
14. Khan, A.L.; Lee, I.-J. Endophytic Penicillium funiculosum LHL06 secretes gibberellin that reprograms Glycine max L. growth during copper stress. *BMC Plant Biol.* **2013**, *13*, 86–100. [CrossRef] [PubMed]
15. Azevedo, R.A.; Gratão, P.L.; Monteiro, C.C.; Carvalho, R.F. What is new in the research on cadmium-induced stress in plants? *Food Energy Secur.* **2012**, *1*, 133–140. [CrossRef]
16. Llamas, A.; Ullrich, C.I.; Sanz, A. Cd2+ effects on transmembrane electrical potential difference, respiration and membrane permeability of rice (*Oryza sativa* L) roots. *Plant Soil* **2000**, *219*, 21–28. [CrossRef]
17. Larbi, A.; Morales, F.; Abadia, A.; Gogorcena, Y.; Lucena, J.J.; Abadia, J. Effects of Cd and Pb in sugar beet plants grown in nutrient solution: Induced Fe deficiency and growth inhibition. *Funct. Plant Biol.* **2002**, *29*, 1453–1464. [CrossRef]
18. Khanmirzaei., A.; Bazargan, K.; Moezzi, A.A.; Richards, B.K.; Shahbazi, K. Single and Sequential Extraction of Cadmium in Some Highly Calcareous Soils of Southwestern Iran. *J. Soil Sci. Plant Nutr.* **2013**, *13*, 153–164. [CrossRef]
19. Greger, M.; Brammer, E.; Lindberg, S.; Larsson, G.; Idestam-almquist, J. Uptake and physiological effects of cadmium in sugar beet (Beta vulgaris) related to mineral provision. *J. Exp. Bot.* **1991**, *42*, 729–737. [CrossRef]
20. Danso Marfo, T.; Datta, R.; Vranová, V.; Ekielski, A. Ecotone Dynamics and Stability from Soil Perspective: Forest-Agriculture Land Transition. *Agriculture* **2019**, *9*, 228. [CrossRef]
21. Marfo, T.D.; Datta, R.; Pathan, S.I.; Vranová, V. Ecotone Dynamics and Stability from Soil Scientific Point of View. *Diversity* **2019**, *11*, 53. [CrossRef]
22. Yadav, G.S.; Datta, R.; Imran Pathan, S.; Lal, R.; Meena, R.S.; Babu, S.; Das, A.; Bhowmik, S.; Datta, M.; Saha, P. Effects of conservation tillage and nutrient management practices on soil fertility and productivity of rice (*Oryza sativa* L.)–rice system in north eastern region of India. *Sustainability* **2017**, *9*, 1816. [CrossRef]
23. Lukin, S.V.; Selyukova, S.V. Ecological Assessment of the Content of Cadmium in Soils and Crops in Southwestern Regions of the Central Chernozemic Zone, Russia. *Eurasian Soil Sci.* **2018**, *51*, 1547–1553. [CrossRef]

24. Burd, G.I.; Dixon, D.G.; Glick, B.R. A plant growth-promoting bacterium that decreases nickel toxicity in seedlings. *Appl. Environ. Microbiol.* **1998**, *64*, 3663–3668. [CrossRef] [PubMed]
25. Zafar-ul-Hye, M.; Shahjahan, A.; Danish, S.; Abid, M.; Qayyum, M.F. Mitigation of cadmium toxicity induced stress in wheat by ACC-deaminase containing PGPR isolated from cadmium polluted wheat rhizosphere. *Pak. J. Bot.* **2018**, *50*, 1727–1734.
26. Danish, S.; Kiran, S.; Fahad, S.; Ahmad, N.; Ali, M.A.; Tahir, F.A.; Rasheed, M.K.; Shahzad, K.; Li, X.; Wang, D.; et al. Alleviation of chromium toxicity in maize by Fe fortification and chromium tolerant ACC deaminase producing plant growth promoting rhizobacteria. *Ecotoxicol. Environ. Saf.* **2019**, *185*, 109706. [CrossRef]
27. Arshad, M.; Frankenberger, W.T.J. *Ethylene: Agricultural Sources and Applications*; Kluwer Academic Publishers: New York, NY, USA, 2002.
28. Penrose, D.M.; Glick, B.R. Enzymes that regulate ethylene levels—1-Aminocyclopropane-1-carboxylic acid (ACC) deaminase, ACC synthase and ACC oxidase. *Indian J. Exp. Biol.* **1997**, *35*, 1–17.
29. Yang, J.; Kloepper, J.W.; Ryu, C.M. Rhizosphere bacteria help plants tolerate abiotic stress. *Trends Plant Sci.* **2009**, *14*, 1–4. [CrossRef]
30. Zafar-ul-Hye, M.; Danish, S.; Abbas, M.; Ahmad, M.; Munir, T.M. ACC deaminase producing PGPR Bacillus amyloliquefaciens and agrobacterium fabrum along with biochar improve wheat productivity under drought stress. *Agronomy* **2019**, *9*, 343. [CrossRef]
31. Danish, S.; Zafar-ul-Hye, M. Co-application of ACC-deaminase producing PGPR and timber-waste biochar improves pigments formation, growth and yield of wheat under drought stress. *Sci. Rep.* **2019**, *9*, 5999. [CrossRef]
32. Jalali, J.; Gaudin, P.; Capiaux, H.; Ammar, E.; Lebeau, T. Isolation and screening of indigenous bacteria from phosphogypsum-contaminated soils for their potential in promoting plant growth and trace elements mobilization. *J. Environ. Manag.* **2020**, *260*, 110063. [CrossRef]
33. Brtnicky, M.; Dokulilova, T.; Holatko, J.; Pecina, V.; Kintl, A.; Latal, O.; Vyhnanek, T.; Prichystalova, J.; Datta, R. Long-Term Effects of Biochar-Based Organic Amendments on Soil Microbial Parameters. *Agronomy* **2019**, *9*, 747. [CrossRef]
34. Ashraf, M.A.; Hussain, I.; Rasheed, R.; Iqbal, M.; Riaz, M.; Arif, M.S. Advances in microbe-assisted reclamation of heavy metal contaminated soils over the last decade: A review. *J. Environ. Manag.* **2017**, *198*, 132–143. [CrossRef] [PubMed]
35. Danish, S.; Zafar-ul-hye, M.; Mohsin, F.; Hussan, M. ACC-deaminase producing plant growth promoting rhizobacteria and biochar mitigate adverse effects of drought stress on maize growth. *PLoS ONE* **2020**, *15*, e0230615. [CrossRef] [PubMed]
36. Danish, S.; Zafar-Ul-Hye, M. Combined role of ACC deaminase producing bacteria and biochar on cereals productivity under drought. *Phyton* **2020**, *89*, 217–227. [CrossRef]
37. Parewa, H.P.; Meena, V.S.; Jain, L.K.; Choudhary, A. Sustainable crop production and soil health management through plant growth-promoting rhizobacteria. In *Role of Rhizospheric Microbes in Soil: Stress Management and Agricultural Sustainability*; Springer: Singapore, 2018; Volume 1, pp. 299–329.
38. Pathan, S.I.; Větrovský, T.; Giagnoni, L.; Datta, R.; Baldrian, P.; Nannipieri, P.; Renella, G. Microbial expression profiles in the rhizosphere of two maize lines differing in N use efficiency. *Plant Soil* **2018**, *433*, 401–413. [CrossRef]
39. Danish, S.; Zafar-Ul-Hye, M.; Hussain, S.; Riaz, M.; Qayyum, M.F. Mitigation of drought stress in maize through inoculation with drought tolerant ACC deaminase containing PGPR under axenic conditions. *Pak. J. Bot.* **2020**, *52*, 49–60. [CrossRef]
40. Zafar-Ul-Hye, M.; Zahra, M.B.; Danish, S.; Abbas, M.; Rehim, A.; Akbar, M.N.; Iftikhar, A.; Gul, M.; Nazir, I.; Abid, M.; et al. Multi-strain inoculation with pgpr producing acc deaminase is more effective than single-strain inoculation to improve wheat (*Triticum aestivum*) growth and yield. *Phyton* **2020**, *89*, 405–413. [CrossRef]
41. Glick, B.; Penrose, D.; Li, J. A Model for the Lowering of Plant Ethylene Concentrations by Plant Growth-promoting Bacteria. *J. Theor. Biol.* **1998**, *190*, 63–68. [CrossRef]

42. Ahmed, N.; Ahsen, S.; Ali, M.A.; Hussain, M.B.; Hussain, S.B.; Rasheed, M.K.; Butt, B.; Irshad, I.; Danish, S. Rhizobacteria and silicon synergy modulates the growth, nutrition and yield of mungbean under saline soil. *Pak. J. Bot.* **2020**, *52*, 9–15. [CrossRef]
43. Miniraj, N.; Prasanna, K.P.; Peter, K.V. Bitter gourd Momordica spp. *Genet. Improv. Veg. Plants* **1993**, 239–246. [CrossRef]
44. Lea Lojkova, V.V. Pavel Formánek, Ida Drápelová, Martin Brtnicky, Rahul Datta Enantiomers of Carbohydrates and Their Role in Ecosystem Interactions: A Review. *Symmetry* **2020**, *12*, 470. [CrossRef]
45. 6 Benefits of Bitter Melon (Bitter Gourd) and Its Extract. Available online: https://www.healthline.com/nutrition/bitter-melon#section8 (accessed on 27 May 2020).
46. GOP. *Fruits, Vegetables and Condiments: Statistics of Pakistan*; Ministry of National Food Security and Research (Economic Wing): Islamabad, Pakistan, 2014.
47. Dworkin, M.; Foster, J.W. Experiments with some microorganisms which utilize ethane and hydrogen. *J. Bacteriol.* **1958**, *75*, 592–603. [CrossRef] [PubMed]
48. Sadiq, A.; Ali, B. Growth and yield enhancement of *Triticum aestivum* L. by rhizobacteria isolated from agronomic plants. *Aust. J. Crop Sci.* **2013**, *7*, 1544–1550.
49. Chapman, H.D.; Pratt, P.F. *Methods of Analysis for Soils, Plants and Water*; University of California, Division of Agricultural Sciences: Berkeley, CA, USA, 1961.
50. Jones, J.B.; WolfH, B.; Mills, H.A. *Plant Analysis Handbook: A Practical Sampling, Preparation, Analysis, and Interpretation Guide*; Micro-Macro Publishing Inc.: Athens, GA, USA, 1991.
51. Nadeem, F.; Ahmad, R.; Rehmani, M.I.A.; Ali, A.; Ahmad, M.; Iqbal, J. Qualitative and Chemical Analysis of Rice Kernel to Time of Application of Phosphorus in Combination with Zinc Under Anaerobic Conditions. *Asian J. Agric. Biol.* **2013**, *1*, 67–75.
52. Bremner, M. Chapter 37: Nitrogen-Total. In *Methods of Soil Analysis: Part 3 Chemical Methods*; American Society of Agronomy: Madison, WI, USA, 1996; pp. 1085–1122.
53. Roth, E.; Mancier, V.; Fabre, B. Adsorption of cadmium on different granulometric soil fractions: Influence of organic matter and temperature. *Geoderma* **2012**, *189–190*, 133–143. [CrossRef]
54. Papoyan, A.; Kochian, L.V. Identification of Thlaspi caerulescens genes that may be involved in heavy metal hyperaccumulation and tolerance. Characterization of a novel heavy metal transporting ATPase. *Plant Physiol.* **2004**, *136*, 3814–3823. [CrossRef]
55. Glick, B.R.; Patten, C.L.; Holguin, G.; Penrose, D.M. *Biochemical and Genetic Mechanisms Used by Plant Growth Promoting Bacteria*; Imperial College Press: London, UK, 1999.
56. Ouzounidou, G.; Ciamporova, M.; Moustakas, M.; Karataglis, S. Responses of Maize (*Zea-mays* L) Plants to Copper Stress.1. Growth, Mineral-Content and Ultrastructure of Roots. *Environ. Exp. Bot.* **1995**, *35*, 167–176. [CrossRef]
57. De Filippis, L.F.; Hampp, R.; Ziegler, H. The effects of sublethal concentrations of zinc, cadmium and mercury on Euglena. *Arch. Microbiol.* **1981**, *128*, 407–411. [CrossRef]
58. Matile, P.; Schellenberg, M.; Vicentini, F. Planta Localization of chlorophyllase in the chloroplast envelope. *Planta* **1997**, *201*, 96–99. [CrossRef]
59. Pankovic, D.; Plesničar, M.; Arsenijević-Maksimović, I.; Petrović, N.; Sakač, Z.; Kastori, R. Effects of nitrogen nutrition on photosynthesis in Cd-treated sunflower plants. *Ann. Bot.* **2000**, *86*, 841–847. [CrossRef]
60. Jalloh, M.A.; Chen, J.; Zhen, F.; Zhang, G. Effect of different N fertilizer forms on antioxidant capacity and grain yield of rice growing under Cd stress. *J. Hazard. Mater.* **2009**, *162*, 1081–1085. [CrossRef] [PubMed]
61. Sharma, S.S.; Dietz, K.J. The Significance of Amino Acids and Amino Acid derived Molecules in Plant Responses and Adaptation to Heavy Metal Stress. *J. Exp. Bot.* **2006**, *57*, 711–726. [CrossRef] [PubMed]
62. Sarwar, N.; Saifullah, S.M.; Malhi, S.S.; Zia, M.H.; Naeem, A.; Bibia, S.; Farida, G. Role of mineral nutrition in minimizing cadmium accumulation by plants. *J. Sci. Food Agric.* **2010**, *90*, 925–937. [CrossRef] [PubMed]
63. Wang, H.; Zhao, S.C.; Liu, R.L.; Zhou, W.; Jin, J.Y. Changes of photosynthetic activities of maize (*Zea mays* L.) seedlings in response to cadmium stress. *Photosynthetica* **2009**, *47*, 277–283. [CrossRef]
64. Shen, W.; Nada, K.; Tachibana, S. Involvement of polyamines in the chilling tolerance of cucumber cultivars. *Plant Physiol.* **2000**, *124*, 431–439. [CrossRef] [PubMed]
65. Datta, R.; Kelkar, A.; Baraniya, D.; Molaei, A.; Moulick, A.; Meena, R.; Formanek, P. Enzymatic degradation of lignin in soil: A review. *Sustainability* **2017**, *9*, 1163. [CrossRef]

66. Datta, R.; Anand, S.; Moulick, A.; Baraniya, D.; Pathan, S.I.; Rejsek, K.; Vranova, V.; Sharma, M.; Sharma, D.; Kelkar, A.; et al. How enzymes are adsorbed on soil solid phase and factors limiting its activity: A Review. *Int. Agrophys.* **2017**, *31*, 287–302. [CrossRef]
67. Tripathi, M.; Munot, H.P.; Shouche, Y.; Meyer, J.M.; Goel, R. Isolation and functional characterization of siderophore-producing lead- and cadmium-resistant Pseudomonas putida KNP9. *Curr. Microbiol.* **2005**, *50*, 233–237. [CrossRef]

© 2020 by the authors. Licensee MDPI, Basel, Switzerland. This article is an open access article distributed under the terms and conditions of the Creative Commons Attribution (CC BY) license (http://creativecommons.org/licenses/by/4.0/).

MDPI
St. Alban-Anlage 66
4052 Basel
Switzerland
Tel. +41 61 683 77 34
Fax +41 61 302 89 18
www.mdpi.com

Environments Editorial Office
E-mail: environments@mdpi.com
www.mdpi.com/journal/environments

www.ingramcontent.com/pod-product-compliance
Lightning Source LLC
LaVergne TN
LVHW070544100526
838202LV00012B/377